MANCHESTER
CITY COUNCIL

CU01454903

Please return / renew this item.

Books can be renewed by phone, internet or Manchester Libraries app.

www.manchester.gov.uk/libraries

Tel: 0161 254 7777

THE EVOLUTION OF MUSIC

PETER TOWNSEND

THE
EVOLUTION
OF MUSIC

through
Culture and Science

OXFORD
UNIVERSITY PRESS

OXFORD
UNIVERSITY PRESS

Great Clarendon Street, Oxford, OX2 6DP,
United Kingdom

Oxford University Press is a department of the University of Oxford.
It furthers the University's objective of excellence in research, scholarship,
and education by publishing worldwide. Oxford is a registered trade mark of
Oxford University Press in the UK and in certain other countries

© Peter Townsend 2020

The moral rights of the author have been asserted

First Edition published in 2020

Impression: 1

Published in the United States of America by Oxford University Press
198 Madison Avenue, New York, NY 10016, United States of America

British Library Cataloguing in Publication Data
Data available

Library of Congress Control Number: 2019941137

ISBN 978–0–19–884840–0

DOI: 10.1093/oso/9780198848400.001.0001

Printed and bound by
CPI Group (UK) Ltd, Croydon, CR0 4YY

ACKNOWLEDGEMENT

During the writing of this book I have benefitted greatly from the critical comments and many constructive suggestions of my musical friend Angela Goodall. It is a great pleasure to acknowledge both her and the many other friends who have made considerable contributions and input to the writing.

CONTENTS

A Note to the Readers   xiii
All the Physics you Need to Know to Read This Book   xv

1. Music—An Ever-Changing Landscape   1
The Universality of Music   1
The Aims of This Book   2
A Rapid View of the Content by Chapter   4
The Standard View of Musical Development   5
The Role of the Renaissance and Reformation   8
My Analogy of Music as a Garden   9
A Difference Between Music and Art   12
The Earliest Signs of Music   12
The Need for Written Records   14
Ritual and Chant Can Enable Technology   15

2. Musical Development Assisted by Technology   17
Western Religious Music in the First Millennium   17
Musical Notation and Printing   19
A Letter to the Corinthians   20
Polyphony in Secular Music   21
More Recent Developments   22
Singing, Secular Music, and the Birth of Opera   23
Survival of Fame and Compositions   24

3. Musicianship and the Cult of Personality   31
The Growing Cult of Superstars   31
The Cult of Conductors   35
Feedback from Critics and the Audience   37
Reasons Why We Can Never Repeat the Musical Past   41
Tempi   41
Musical Training   44

Changing Styles of Performance   46

Who are musicians?   47

The Relevance of a Changing Landscape for Earlier Music   49

4. Signal Processing by the Brain   51

Sensing and Survival   51

Alternatives of Discussing Wavelength or Frequency   51

Sensing Light   52

A Sound Strategy   54

The Role of the Brain   57

The Design Challenge for the Ear   59

The Frequency Dependence of the Ear   66

Hearing Quality from Historical Records, Damage, and Aging   69

Hearing Loss, Hearing Aids, and Their Relevance to Music   71

5. Interpreting Complex Sounds   75

The Brain and Signal Interpretation   75

The Harmonic Content of Notes from an Isolated Violin String   76

The Role of the Violin Structure   79

Brain Tricks to Guess at a Fundamental   80

Anharmonic Signals   81

The Strain on the Brain   83

A Final Comment on Hearing   84

6. Scales—Idealism or Compromise?   87

*Musical Scales—the Conflict Between Idealism and
Compromise*   87

Pythagoras and Simple Musical Intervals   89

Pleasant Musical Intervals According to Pythagoras   93

Wavelength and Frequency   94

Later Ideas on Scale Tuning   94

Summary of Ideas on Defining Scales   96

Does Equal Temperament Tuning Really Work?   97

Naming the Notes   99

Final Thoughts on Scales   100

7. Musical Changes Driven by Technology   103

Why is Technology Relevant to Music?   103

The Violin Family   105

Violas, Cellos, and Basses   109
Improving Wind Instruments   112
Examples of Instrument Evolution   116
The Clarinet Family   117
The Saxophone   118

8. Development of Fixed Frequency Keyboards   121
Keyboard Instruments   121
Clavichord and Harpsichord   122
Piano Development   124
Dilemmas with Piano Music   126
Electronic Pianos   129
Electronic Instruments   132
Other Technological Influences on Music   133

9. Decay of Information and Data Loss   135
Technological Driven Changes in Music and
Composition   135
Survival of Literature and Painting   136
Film and Video Survival   139
The BBC Doomsday Project of 1986   140
Technology and Recording   141
The Recording of Musical Sounds   142
Music Scores as We Have Known Them   144
Will Printed Scores Survive?   144
Embellishment and Improvisation   147
An Overview on Loss of Musical Information   148

10. From Live Music to Electronic Offerings   151
A Century of Electronics and Music   151
Technology and Music Processing Over the Last 100 Years   153
Microphones and Loudspeakers   154
The Alternative Scenarios   156
Sound Engineers, Live and Record Mastering   157
Directors and Camera Crews   158
Can Recordings Give a Faithful Rendition?   160
Hidden Factors in Record Making and Mastering   161
Future Electronic Screen Displays   164

11. Analogue or Digital Recording   167
    Vinyl   167
    Magnetic Tape Storage   169
    Experimentation with Tape for Popular Music   171
    Advent of Compact Disc Recording   172
    Local or Cloud Storage   176
    Where and How Do We Listen?   177
    The Future   180
    Downloads and Streaming—Will They Change What We Listen to? 182

12. The Voice   187
    The Intrinsic Music of Humanity   187
    Separation of Song, Dialect, and Language   190
    Voice Training and Singing Teachers   192
    Tone Quality and Voice Control   196
    Historical Perspective of Frequency Analysis   198
    Future Developments   199

13. Acoustics of Concert Halls and Rooms   201
    The Sound of Music   201
    Sound Paths from Stage to Seat   203
    Reverberation and Echoes   205
    How Meaningful is Citing a Reverberation Time?   207
    Controlling the Reverberation Time   209
    Experience in Improving Acoustics   212
    How to Mix Soloists and Orchestra for an Opera   215
    Concert Hall Reflectors, Clouds, and Sound Diffusers   217
    Listening to Music at Home or in a Car   218

14. Orchestral Layout and the Best Concert Seats   221
    Orchestral Size   221
    Time Spreads Across a Big Orchestra   222
    Identification by Direct Sound   223
    Analysis of the Spectrum of Component Notes   224
    Factors that Change the Sound of Instruments   225
    Directionality of Instruments   228
    Arrangements of the Orchestra   230
    Where Are the Best Seats?   234

Musical Changes with Distance   237
The Final View on Choosing a Seat   239

15. Music, Emotions and Political Influences   241

Introduction   241
Funding and Power Bases of Music   242
Responses to Music   243
Psychology and Musical Impact   245
Political Influence and Control   248
Is a Musical Profession Free from Politics?   251
Earlier Dissemination of Music   253
The Use of Music for Political Control   254
Music in Religious Contexts   257
Responses to Regional Variations   259
What Constitutes Music that We Like   259

Further Reading   263
A Brief Glossary of Less Familiar Terms   265

A NOTE TO THE READERS

Music is universal, and we do not need musical training to enjoy it. It has never been static, and the form, complexity, and styles not only differ across the globe but it has existed since prehistory. Music has evolved with time, and benefitted from advances in many sciences, as well as changes in social conditions. Music and technology are totally intertwined, interactive, and mutually beneficial. New instruments, materials, and understanding of acoustics have changed both musical styles and musical cultures. Equally, there has been intense feedback to improve the science, and we may not immediately realize that it was music that forged the advances in sound recording that produced microphones and amplifiers. Music is thus the parent of all the electronic technologies of the modern world. My aim is to offer thoughts on how music has benefitted and developed from such advances and innovation in the development of instrument design, concert halls, and electronics.

Music is an incredibly powerful emotive force which has been exploited throughout history to influence people in terms of religion and politics for both nationalism and warfare, as well as often defining social class divisions. What we mostly think of as an activity for pure pleasure is in fact far more subtle and a deeply entrenched feature of our lives.

In explaining music from early origins to our current involvement with performance, broadcasting, and recordings, many links with technology will emerge. However, I have consciously aimed at a broad audience who do not need either musical or scientific backgrounds, and recognize that the interaction between science and music is actively changing across all musical genres. In addition to the main text, there are short glossaries of musical and less familiar technical phrases that appear in the book.

My thoughts on the role and interplay between technology and music are distinctly personal and my opinions will sometimes differ from other commentators but I see this as a stimulating bonus. I am a scientist by profession and am often delighted to realize that people who see themselves as non-scientists invariably have a far better intuitive grasp of science than they realized. My non-science friends who have read the drafts have been surprised that they have had no problems following my views on the interaction between science, musical appreciation, and composition. Knowledge can equally enhance our enjoyment of music, and I, at least am wiser and more musically appreciative as a result of writing this book.

I have bravely made predictions as to future directions, highly dangerous as it is incredibly easy to be wrong, so if your predictions are better than mine do not be amazed, and I congratulate you in advance.

To explain my interest in both music and technology, I can say that both are central to my life. I started with the violin at the age of 4, and have played in orchestras and quartets, taught courses on the physics of music, and researched the design of violin-making. I also enjoy singing. Scientifically my work has been multinational and spanned many disciplines. Examples of topics range from technological applications of imperfections in materials to novel products, advanced lasers, optical biopsy for cancer detection, radiation dosimetry, archaeology, and geology. This eclectic and wide span of activities has helped me to appreciate the diversity of background experience that we bring to a subject as varied as the umbrella we call music.

ALL THE PHYSICS YOU NEED TO KNOW TO READ THIS BOOK

Physics textbooks on sound have been around for nearly two centuries, so we have a fairly good understanding of the basics. The essence of the topic is that if we make the air pressure oscillate this will set up a wave that travels away from the source. Once this oscillating pressure reaches our ears the ear drum vibrates and couples the energy into the inner ear. Our biology produces a tiny electrical signal that is fed into the brain, and this then makes fantastic efforts to interpret what the sound is all about. That is it!

Writing a scientific book about sound needs a bit more to make it marketable, but here I am just interested in appreciating music and trying to see why it is evolving. So we do not need the fine details. There is an immediate link between the science and music, as when a simple metal tuning fork vibrates, the pressure oscillates at a frequency defined by the size of the tuning fork. In this very basic example the amplitude changes smoothly with time (and has the form of a sine wave). We describe this repeating pattern in terms of a wavelength or frequency. The two are linked as the product of the wavelength and the frequency defines the velocity of the sound as it travels towards us. This is precisely the equivalent of finding our walking speed, where one complete left plus right pace is the wavelength, the number per second is the frequency, and when we multiply the two together we have defined our speed. Sound travels at ~330 m/sec in air, but ~12 times faster in a denser material such as a wooden floor or ~18 times faster in a bar of steel.

Except for the tuning fork nothing else in music is that simple. Violins produce pressure waves that are not smooth sine-shaped ripples, but instead are more sawtooth in shape. These are easily analysed by mathematicians into a related set of overlapping waves (e.g. as in Chapter 5). The same signal processing is incredibly simple for the ear/brain system, but a more complex challenge is set by the voice and singing. Chapter 12 will emphasize the complexity and mixture of different frequencies that are formed when we sing an apparently simple single vocal note and vary the tone of the sound.

The truly impressive brain power involved in listening to music is that we will have many sources from a choir and symphony orchestra, often totalling more than a hundred individuals, and different musical parts. Despite this complexity we can analyse the overall picture and separate different parts and patterns in the words and the music. Precisely the same skill is used to listen to an individual speaking in a crowded noisy environment.

You already possess this superlative analytical ear/brain tool for signal processing and so you are automatically an expert in the physics needed to enjoy music of every age and genre, and, I hope, this book.

CHAPTER 1

MUSIC—AN EVER-CHANGING LANDSCAPE

The Universality of Music

Music is international, multi-cultural, and deeply embedded in the human psyche. Yet our understanding and appreciation of it is totally personal. Even our musical responses, tastes, and appreciation are never constant, but change continuously with time as the result of all our experiences. Musically we are influenced by everything we have ever heard, including sounds from new sources, as well as the result of repetition of the same items, and interpretations from different performers. Recordings are so accessible that we can hear the same works over and over again, a situation that was impossible just one or two generations ago. This may be progress, in terms of ease of access, but it may equally dull our concentration and appreciation. Trying to understand such a dynamic situation, and to consider if there are any clear patterns in the way we develop our enjoyment and understanding of music, is a difficult challenge.

One example which emphasizes just how basic is our musical conditioning is that the music we hear while still in the womb has an influence on our future enjoyment and preferences for musical styles. This is probably not too surprising as the sounds we heard in the womb, such as our parents' voices, were essential to help imprint our recognition of them once we emerged into the unprotected world. Nevertheless, this is a gem of information if we want a musical child, as we can influence the sensitivity and choice of music style by playing our preferred music to it in these crucial natal nine months. Not only is this useful information if we wish to raise a child with interest in a particular type of music, but such conditioning dramatically reveals the ability of music to influence us, and therefore it emphasizes how deep-seated is our human dependence on music.

As we are growing up, our enjoyment, or dislike, of music is modified by our culture, our abilities, knowledge, nurture, and our previous musical experiences, as well as what we are currently hearing. Fortunately, we are not predestined to appreciate just one style of music, nor are we limited to one regional type. Many people have a moment of revelation when they hear a particular piece in an unfamiliar style and then embrace such music, and so broaden their horizons. These key step changes can equally be between 'classical', jazz, current pop styles, or even music from other cultures. Diversity in taste can offer great rewards.

Once we realize that we have the options to make changes, whether at 16 or 61, we can consciously influence our progress and enhance the pleasure. It is equally helpful to realize that, contrary to many claims, no one type of music is superior to another. There are just differences; although within each particular genre there will be performers who are more skilled or sensitive presenters of that style. Since we are not confined to enjoying just one musical style there is no limitation set by listening to classical music, and both adults and children should ignore pressures exerted by colleagues or school friends to only play fashionable music. This is often the case with children and pop music. As in other aspects of life, a degree in chemistry or physics does not inhibit us from reading fiction, languages, or philosophy, or writing poetry. We exploit just a fraction of our brain potential, and adding diversity to our activities enhances all of areas that interest us.

Musical influence, and preferences, will vary throughout our life, and we may take comfort in the observation that elderly people who suffer from variants of memory loss will invariably retain musical enjoyment, and be able to recall songs and their words, even if they do not recognize their own relatives. Even more positive is that music is well-documented to be a successful therapy for the elderly, and it extends mental activity.

The Aims of This Book

I have several related aims within the scope of this book. The first is to see if there is a pattern in the way in which music has evolved, and is evolving. Since it obviously is doing so, I also want to understand how this influences our appreciation and performance of music from different historical periods. The timescale is not just for works written in the past 200 years, but applies equally to why, and how, our taste and performance standards are changing on the timescales of one or two decades. If we ask a rather trite and simplistic question such as 'Can we ever hear or play the music of Beethoven in the way it was first performed?', my answer is going to be a very definite 'No', as not only have the instruments and concert halls changed, but our musical and cultural backgrounds, and expectations, are incredibly different. I will attempt to justify such an emphatic position, but will do so with the happy realization that most people will probably gain more pleasure from the music now than was possible under the conditions when such works were first performed. The diversity of musical styles has also become available to a much wider spectrum of all the social classes.

The second broad objective is to indicate how science and technology have played a major role in the changing design of features as diverse as musical scales, construction of instruments, and concert halls. These non-musical changes directly influence both our enjoyment and the compositions appearing from composers of each generation. Inventions of new instruments is continuous and the new toys have instantly been exploited by both performers and composers. There has been an equally dramatic technological input that has emerged through recording, broadcasting, and concert hall amplification and, particularly because of the advances in electronics and recording techniques, so that we now have constant access to music from skilled performers.

Radio, vinyl discs, CD, MP3, MP4, and other versions of the music differ from live events because of differences between the acoustics of concert halls and home listening. Nevertheless, they colour our expectations of how music should sound. This is not automatically a bonus because the balance between instruments and between performers is generally quite different on a recording compared with listening to a live performance. One obvious example occurs with church organ music, which in the ambience of a large echoic building may not just be heard, but physically sensed, as the powerful low notes can be physically felt. The same performance heard in a normal room will lose all this stimulation and excitement. Further, the sounds are often 'muddy' and blurred because our home listening does not benefit from the long reverberation times, as music echoes around large church buildings.

There are differences in viewpoint as to which might be preferable, but live concert, broadcast, and recorded versions are best seen as complementary. The variations in each (even in the choice of concert hall and seat) may be far greater than we would initially guess. I will offer some surprising comments on this topic! Our own inputs and responses are equally crucial, and certainly the level of concentration we put into each piece can be very different, and is never the same. In reality, this means we never hear an identical performance, even from a CD.

The electronic impact on music, during both composition and performance, has scarcely been felt compared with the potential changes one can envisage. I will make some predictions as to where this might lead in the future. As a scientist, my instinct is to say that I would hope that the developments driven by technology would be beneficial. But within most technologies we first focus on the benefits, but invariably we fail to recognize that there can be downsides. This is equally true for music, and I will comment on some major potential problems. I am sure that in 100 years from now, music, and the technologies of the way we play, present, and record the sounds, will be greatly different from those of today.

A Rapid View of the Content by Chapter

The chapter headings will track that we are in a continuously changing situation (Chapter 1), which is strongly defined, directed, and modified by technology (Chapter 2). In Chapter 3, I consider how musical taste is often a separate issue, and it can be easily influenced by the emergence of superstars and virtuosi. The two may, or may not, be synonymous. Certainly, in our own generation there are many examples where status, and image, are driven by profits and commercial hype from recording companies rather than ability.

Musical training and experience are very personal and difficult to quantify, but other features which have helped in the developments in music are much more tangible, and we can list and detail them. Our individual input is defined by the signal reception and processing with ears and brain (Chapters 4 and 5). There is the pragmatic fact that music uses a wide range of scales across the world (Chapter 6) which colour our preferences for music. More detailed examples of where technology has made major impact on instruments are considered in Chapters 7 and 8. Life is not all progress, and both data and knowledge are easily lost or fade away, so this is discussed in Chapter 9. Technological progress is equally transient and changing rapidly. This has driven many developments in both electronic recordings and music (Chapters 10 and 11). Singing is a basic human activity, so I have included comments in Chapter 12 on how electronics can help in voice training and of course can help with electronic hearing aids. Finally, in terms of technology, there has been an immense, and valuable, impact from the sciences of acoustics and architecture, as sketched in Chapters 13 and 14. The final Chapter 15 looks into the realm of the psychological and physiological impact of music on our lives. It is deep-seated and powerful, but is so much part of life that we may not even recognize how it is used, and what it can achieve.

My progression of chapters has some logic, but I have tried to write the chapters so they could be skipped, or taken out of sequence, if that is your preference.

The Standard View of Musical Development

Many musicologists have written about the historical developments over the past few thousand years. The details for the earliest types of music, and how and why they changed, are often just intelligent guesswork, but the progressions over the past few hundred years are fairly well documented. They have a direct bearing on 21st century musical taste and appreciation. As one who is used to hearing European music, such writings on the older European work are generally helpful, as there have clearly been some patterns emerging that resulted from cultural, religious, or other historic influences. Perhaps surprisingly, because the thoughts are often in the realm of philosophy and opinion, there is a broad agreement in the assessment of key factors. I suspect this is because the earlier writers came from similar social backgrounds, and there was a tendency to quote earlier references. Modern writing is fortunately far more diverse and includes input from a wider range of people.

The lack of significantly different interpretations of musical development before say 1500 AD will reflect the fact that change and progress was very slow and, at least in European culture, was totally dominated by the church. By about 1500, the tight control of religion had weakened, and new ideas were emerging in literature and art, as well as some science, and all these factors were inevitably reflected in less stringent control of church music. That was also the period of the Reformation with schisms in religious beliefs, and again this led to diversity in both church and secular music.

I will capitalize on the existing writing about early musical history, and rapidly mention some of those cultural developments that have helped to shape our Western music. My only hesitation is that such perceived patterns are often rather philosophical views that are hard to quantify and substantiate, and the literature also has a weakness in that there are many cross references to a rather limited group of prime authors. Such a pattern does not constitute proof, but it indicates that there was a mainstream view on such changes. Certainly, not all factors have been included, and from the background literature one sees that very many articles and books are just centred on developments in Western music, particularly at the classical end of the musical spectrum. There are often only token references to musical developments in other cultures. This is unfortunate as the modern access to worldwide music needs to include a far broader perspective. Indeed, looking at the membership of current Western orchestras it is clear that many of the soloists, conductors, and players are contributing musical, ethnic, and cultural backgrounds with far greater diversity than existed even 30 years ago. Their influences will appear in the music.

Musicians in many countries with European links, such as the Americas, have strongly advocated developing national styles which are not repetitions of European music. These changes are global versions of the way Europe progressed in earlier centuries when there were fashions for, and against, French, German, Italian or other styles of compositions and playing techniques.

The standard summaries of Western culture use approaches that attempt to link musical evolution with increasing diversity and experimentation in the arts and literature of the various periods. Undoubtedly an overall cultural atmosphere which allows new ideas and opinions to flourish will be equally flexible with musical or scientific ideas. The reverse is also true, and in the 20th century the strictly applied state control in say Russia or Germany, had a very obvious effect on the styles of music that could be composed and performed. Political constraints imply economic constraints, which in turn influence and stifle the ability of musicians to perform and compose, because they need audiences and funding to further their creativity. Basically, for most composers, no money means no musical output. So, when the Western world was dominated by the church, then religious music was especially favoured, and predominant. Equally, nationalistic doctrines can cause a temporary stagnation.

Folk music is highly varied in objective, from simple pleasure, to work songs, country fairs, working in fields, co-ordinating sailors raising sails, and marching songs for the army. Music has a role in legends and history as well as ballads, so there has always been a wealth of opportunities for a mass market, and it had a separate and freer existence, being to some extent less controlled than music that was designed for religious purposes. The secular examples have run in parallel with 'official' music and, because the performers were frequently amateurs, not professional musicians, the songs, dances, ballads, and story lines needed to appeal to a wide audience. It also meant that the songs had to be tuneful and have mass appeal, so they could be remembered. This implies that they were of freer formats than those in more formal church items.

There are many examples of folk music that was transmitted entirely aurally, and for example in the Caribbean there are still small communities who sing and dance music that came from Africa two centuries earlier, when they became enslaved. None of this music has been written down, so it is vulnerable if the community disperses or dies out. Such heritage music and rhythms may differ considerably from other types of songs and dances.

The social divisions of music, and those writing about it, have often resulted in the popular folk music being underrated in the past. This sense that 'classical' is better than 'popular or folk' was epitomized in the first 30 years or so of the BBC, where there was great reluctance to have any music that was not pure establishment, with a nominally intellectual or educational content. Even the 'light classic' programmes such as the Grand Hotel Palm Court barely survived, despite the fact that such a programme had listening audiences that were more than ten times greater than for any other music programme. This is in complete contrast to the modern world where 'classical' music concerts and CD sales are very much a minority part of the overall wider musical scene. It is a salutary comment that vinyl disc sales by the Beatles exceeded the *total* of all the worldwide vinyl sales of classical music!

Further, attendance at 'classical' music concerts strongly suggests that there is a generational divide in which grey hair is a prerequisite for being in the audience. To me it is unclear if this is caused by a real switch in musical interest, a total lack of music teaching in many schools, or a younger generation who want instant communication, and cannot maintain interest in works for more than five minutes. There can also be peer pressure in school and teenage communities, that 'classical' is not cool. I know of examples of skilled classical music teenagers who have given up, precisely because of peer group pressure. Attendance at many opera houses may understandably be limited to the grey-haired wealthy, as the price of the seats is invariably extremely high. To help with the high production costs, some opera events are now simultaneously relayed to perhaps a thousand cinemas, or on TV. The seat cost is much lower, and the close-ups far better than one can see in the live performance. However, as always, there is a trade-off between the electronically modified acoustics, and the overview of the actions on stage. Financially, this may not be as rewarding as one might imagine, at least back in 2010, when the Metropolitan Opera in New York only gained ~$11 million to offset annual production costs of $325 million. However, the practice is now far more fashionable. Cinemas gain particular benefit as the performance times are often early afternoon, when there are normally few customers. In the UK, cinema income from this source rose from almost zero in 2009 to more than £40 million by 2014.

The Role of the Renaissance and Reformation

Before looking at pre-history or the music of ancient civilizations I will start at a time when there were manuscripts and comments on Europe in the Middle Ages. For us in the 21st century, it is extremely difficult to imagine life in Europe in the Middle Ages, but without doing so we cannot put their music into perspective. In the period from say 1400 to 1500 AD most European capitals and major cities had populations of well under 100,000 people. There was an immense divide between the few in the rich or ruling classes, and the bulk of the population. Crucially, the people were deeply religious and superstitious. While the favoured classes had time to pursue or patronize musical activities, the overall pattern of music was strongly defined by the church, and any music that might be considered joyful was in danger of being considered sinful. So this was not an ideal time for musical innovation, especially in church music. Church control of music, science, or even the colours that could be used in painting, were evident and enforced. For a city such as London, with population of under 80,000 (including women and children), one may guess that the number of professional musicians and composers would have been no more than a few hundred men (women were less likely to be composers at that stage). By contrast, modern London has at least 8 million people with musical input from both men and women, from all social classes, and with a far greater adult life expectancy in which they can be musically productive. London is now also more varied, as estimates suggest that more than 30 per cent have grandparents from other nations across the world. Musical and cultural diversity has thus flowed into this generation.

However, during the latter part of the Middle Ages there was a steady pressure to challenge the church control, and innovation crept in and changes were accepted in aspects of painting, literature, science, and music. It was not a true break from religious controls, but it allowed a wider diversity of musical composition. Many people have recognized that one key aspect of the cultural change was to shift the emphasis from all things being centred and controlled by religion, to a development which allowed some individuality. This trend evolved into the now familiar cult of the personality. Musical compositions were allowed to be attributed to specific composers, not some unnamed monk (Anon was a very prolific composer). There was experimentation with new instruments, and tonal patterns, chords, or singing in parts, rather than the single lines we think of as early monastic church music. Another critical key technological advance was the introduction of musical notation that could be written down, and together with the printing press, it meant that music could be disseminated in bulk, not just by hand-copying. Here was a real bonus of a technological input.

The broad term of a Renaissance, which certainly occurred in art and literature, is usually cited as driving the progress in musical styles. This is partially true, but the more obvious effect for music may be the growth of emphasis on an individual performer. This has steadily moved to the present day with the acceptance of superstars in terms of composers, musicians, and conductors, as well as high-profile personalities in TV, film, or mass media that we seem to crave nowadays. A sociologist might ponder whether our focus on superstars is another version of Greeks and Romans inventing the stories about their many superhero gods.

My Analogy of Music as a Garden

In many ways, I feel the standard imagery of linking music to art and literature is quite limiting. Instead, I would prefer an analogy between music and gardening. Music and production of vegetables are essential foods for the mind and body, and the flowers add perfume, beauty, and intangible escape from the realities of daily life. The things we grow are increased by the introduction of new products from more distant countries (maybe mundane potatoes, or exotic spices) and their form and taste are evolved by our selective retention of particular variants. To enjoy the garden, we do not need to know the names of the plants or where they came from (unlike the interpretation of paintings or literature). Music fits this pattern perfectly, and even today we see classical music being revised by sounds of jazz, South American dance, Indian instruments, or African rhythms etc. Survival of earlier music has had the same fate as early plants. Good apple varieties are often displaced by supermarket forces, even though they had taste and texture that may now be lacking. Vegetables are often bug free, not because they are better, but because no self-respecting bug would eat them. There are parallels between the former garden plants and music, as modern musicologists are trying to popularize earlier works. With an improved understanding of genetics, modern biologists are attempting to preserve seed banks, and have repositories of rarer plants to minimize total extinction. Recording systems are doing the same for less popular music, both from ancient to modern works.

The compositions of many prolific composers have been almost totally lost by changes in public fashion. Survival of others has happened via revivalist crusades by a very few individuals. Compositions by people such as Corelli, Bach, or Monteverdi (who were eminent in their day) faded from the repertoire, and were only reintroduced through good publicity. Their work has re-entered the lists of popular musicians. However, for many other composers, who may have had greater fame and prominence in the past, their work is now hidden in the shadows, or totally lost.

I also like the gardening analogy because it has a far more universal perspective than comparisons with art or literature. Crucially, it is language independent. Operatic arias and songs can be enjoyed quite independently of our knowledge of the language being used, just as we may enjoy food or perfumes without knowing the name. Indeed, I have often felt that once one understands the text of many operas, then the weaknesses of the libretti, or the trivialities of the words, means that knowing the language could actually detract from the pleasure of the sound. Such a comment needs clarification, as when *listening* to operatic music I find the music is crucial and the language unimportant. However, if I am actually watching an opera I need the background plot, and then the super-title translation of the text is valuable. Similarly, silently watching a foreign film with subtitles is not enough, and I still need the sound track, even if I understand only a few of the spoken words. The music of speech conveys far more emotive content than written subtext words.

My garden flower version of this is that cut flowers in the house can be admired for their colour, form, and beauty, but in the garden, one must see them in the context of the other plants, and the perfume of the flowers is much more important to make them stand out from the crowd. In both cases, we are benefitting from the standard human response that more than one type of simultaneous stimulation (sound, sight, smell, or touch) of an event, enhances our overall sensitivity. The downside of this effect is that listening to recorded music is potentially less exciting than hearing a live performance, and we can readily be distracted by other activities.

This is very obvious when I go to local concerts by people who are not professional performers. They are typically good musicians, but not world class. However, such local live events are more than just enjoyable and satisfying, as there is the magic of immediate presence with the performance. Often the theatres tend to be small and intimate, and when sitting directly next to a floor-level stage at an opera such as La Bohème, it is incredibly emotionally moving, at a level one rarely experiences at long range, separated from the stage, in a full-size opera house. The same bonus emerges when comparing the live instrumental concerts with top flight performers on a CD with home acoustics.

This need for actual participation, or presence at an event, is equally true in most aspects of life. Millions will sit at home and watch football in the warmth and comfort of their homes. They are not actually watching the total game, just the limited view and highlights selected by the camera team. Going to the stadium, in the cold and rain may be more expensive, but it gives the total perspective of the play, plus the high emotion generated by all the other spectators. The only higher level is to be one of the players. I know from my own experience that competing (in my case fencing) is emotionally satisfying (even when losing), and watching is not. Similarly, in a string quartet the best seat for total involvement is when playing with the other three.

A Difference Between Music and Art

Many music historians draw close analogies between the changes in art and music. There are some, but comparisons of music with art are weak, because a painting is static and fixed in the time, format, and perspective or the painting techniques in use when it was created. Not only is it locked in time but viewing it may well depend on a great deal of historic knowledge to recognize the symbolism, mythology, allegorical or other identifying signs, and artefacts shown on the pictures. There is also the dilemma that, without a catalogue or description, we will not know who the people are meant to portray. By contrast, we can interpret music as we wish, and, if we choose, modernize the performance through new interpretations, even for a 400-year-old composition. We are not restricted and can play works with different instruments. Reference to any modern CD catalogue will show just how many performers think they can provide a new, or better, take on classical favourites. My comparison with a garden is that we can find beauty in the flowers from shape, colour, and perfume, and it is never static. Plants grow and change as does our musical appreciation. New features continually come and go and there will be weeds or mistakes. Overall, the appearance depends on the soil, the layout, and the gardener. Even if many people praise the garden, then no two will ever see it from the same perspective, or detect the subtle perfumes in the same way, nor will we ever see the same garden in an identical way on successive visits. For me, such variations offer the pleasure of a garden, and are exactly the same factors which provide the excitement and individuality of musical performance and listening.

The Earliest Signs of Music

I will briefly follow the standard pattern of describing musical development from pre-history to say the 17th century. From there forward, we will be solidly into the music which routinely appears in concerts, and on radio and TV broadcasts. This is not to say earlier music is not performed (and enjoyable), but only that there are some real differences in why it was written, and the intended audiences for the music.

Once we became articulate humans and listened to the songs of birds, and the other sounds in our world, it seems inevitable that we would try to emulate some of them, and our attempts would lead into singing. Archaeologists have found bone-flutes dating from before the last ice age. In at least one example, found in southern Germany, a flute has been carbon dated to more than 30,000 years old. This flute, and related artefacts, may not have been only for music, but perhaps served as bird lures. Nevertheless, they show an initial excursion into musical instrument making, because this flute, with four holes, was clearly quite a sophisticated item to manufacture, needing both musical and mechanical skills.

Archaeologists have identified other items they think had musical usage, from as early as the Palaeolithic, and also from Neolithic times. These have been discovered in many parts of the world. Other obvious basic instruments that were easy to construct include clapping sticks, gourds, and drums. Variations in drum size indicate that several notes were in use, and this is the same pattern as has been noted in current tribal societies. In terms of simple instruments standard guesses are that a plucked string, blowing into reeds and hollow tubing, or banging on pieces of metal would all have taken us on our first musical steps. All these have appeared in carvings from earlier civilizations across Africa and Asia. There are cave paintings and carvings, with such instruments being in use for at least some thousands of years. The much later Egyptian cultures have clear images of musical performance for pleasure with instruments recognizable as variants of items such as harps, drums, and lutes. Carvings, as in Assyrian examples, have trumpet style instruments in battle scenes. According to the book of Daniel in the Bible, in Babylon, King Nebuchadnezzar had a royal band with brass, wind, and string instrument, together with drums. Religious images from many faiths across the world have similar identifiable instruments.

The images on carvings and paintings probably refer to specific events, so the instruments were being used as status items for special occasions. Other instruments existed for non-ceremonial purposes, and ancient Greek culture routinely employed musical entertainers for social events for the men at home. The entertainers were usually women, in the case of the Greeks. Such an historical pattern is not confined to the Western world and equally clear examples of closely related variants of these instruments and usage appear from Chinese to South American carvings and paintings.

There seems to have been parallel inventions of drums, flutes, and harps in civilizations as separated as China, India, and Egypt some 3000 years go. There is no suggestion that these early instruments were copied from a basic source, rather, the images show that there were parallel developments throughout the world. The same is true of the progression from stone tools to copper, bronze, and iron artefacts, and unfortunately, the universal development of weapons. These parallels across the many emerging civilizations of the world emphasize that human development is hard-wired to include music as an essential part of being human.

While we have no idea how these early instruments sounded, or indeed the pattern of the music, the various images suggest that they could be played either in isolation, for example a singer with a lute style instrument, or in a consort, as in religious or military groupings. This would strongly suggest that they were not merely played in unison, but that part playing and chords were in use. This view is distinctly contrary to some music texts that suggest that only a single line was played until the use of parallel, or different sounds and tunes (called polyphony) appeared in European music. I suspect that this Eurocentric view is just a subconscious attempt to promote the local culture as superior to anything 'foreign'.

The Need for Written Records

Music lagged behind language in that it would have still been orally transmitted long after the various attempts to write speech with lines in clay, hieroglyphs, pictograms, and alphabets. This lack of a written 'language' for music may now seem problematic to us, but realistically, even in Europe, literacy was not universal, and so speech, stories, and music would have survived by repetition between generations. Music is not a separate or unique part of human life. For a non-literate society, there are various ways in which we can reinforce the effectiveness of passing on information to the next generation. Songs or chants were, and are, a universal route for this. We are certainly a highly favoured species and reasonably intelligent and, to a very large extent, our rise from caveman status is because we have the ability to communicate by speech. A multi-sense approach with words, images, writing, and music is therefore particularly effective.

I will return to this theme with more examples when discussing the loss and decay of information and knowledge.

Ritual and Chant Can Enable Technology

This is a general feature of all different civilizations and cultures. For skills which involve a complex applied technology the musical route may still be helpful. For example, in the construction of the very time-consuming process of making Japanese Samurai swords, there are a whole range of metallurgical steps. These begin with the methods of purifying the starting materials, then continue with ways to heat-treat the iron, and separately add different amounts of impurity in the outer skin, and the central core of the blade, before bending, heating, and hammering the two parts together. Depending on the concentration, the use of carbon impurities in the iron alloy can either forge a strong core blade, which can resist shock, or an outer skin which can be sharpened. Song and chants can help with the detailed remembering of the complexity of this processing. Hence, the approach used was to formulate the work as a religious ritual, so that all the key steps were retained. The steel example underlines that we can make progress without a written history of the techniques, even for technology that was literally 'cutting edge'.

The Japanese sword example has a final finishing stage that results in it curving and displaying a wavy pattern along the sharpened edge. Most texts assume that this was a unique invention in Japan, However, in the writings of the Beowulf sagas, there are similar descriptions of swords with a snake-like marking along the edge. So, two widely separated, and disparate, cultures had achieved the same invention. We therefore should not be surprised that the more easily fabricated musical instruments have also appeared independently in many cultures.

The negative side of speech and musical transmission of information is that material is easily lost if for any reason there is a hiatus in the process (e.g. a war, plague, or fashion). Similarly, the music and knowledge that do survive can be modified by error, or intent, during successive generations. There are still many parts of the world where music is played, but the performers are illiterate (or cannot write music). There are some clear benefits in this tradition as many types of music survive with a non-written tradition. The ongoing examples of inventiveness in jazz etc. depend on the skill of the performers. Hearing the same players more than once does not mean that we will ever hear the same version of a familiar basic tune.

As listeners, we are similarly performance dependent, and not just a passive part of the musical experience. We hear and appreciate different aspects depending on our mood, and concentration, as well as the ambient conditions of our listening. The same CD can vary between mundane and deeply moving, depending on our current frame of mind and level of attention. For example, concentrated listening to a music CD with headphones that remove background noises, hearing it while typing, or when driving in a car, all effectively give different views of the highlights. These examples show a true involvement in the language of the music, in the same way that we would never expect to have an identical conversation with someone, even if we are re-discussing the same topic.

By a few thousand BC, written languages had appeared, but this was not immediately followed by a method of recording musical notation. Nevertheless, music was essential at the highest levels of many civilizations. In China, music was highly rated and the 'Golden Emperor' of ~2700 BC wrote a long-term popular item, and the work was praised much later by Confucius around 2250 BC. Undoubtedly the music was not trivial, as in the same period a Chinese scholar, Ling Lun, had not only discussed their musical scale, but added names to the notes. China, in common with most areas of the world from the Pacific to Africa, Asia, North American Inuits, as well as the European Celts, had all arrived at a five-note scale, which we now term pentatonic. Ling Lun gave names to the various notes in terms of social status. The names ran through kong (emperor), chang (minister), kyo (senior merchant), tchi (official) and yu (peasant). In Western music, we have moved away from the pentatonic scale, but an easy route to approximate one variant of it is to play the five black notes on a piano keyboard. Starting on F sharp on the piano these define a pentatonic scale (F sharp is the left key of the group of three). This piano spacing of musical intervals is just one of many pentatonic scales that vary with country and region. To a native user of a pentatonic scale, the Western scale probably sounds as exotic as the other pentatonic ones do to us.

I will discuss musical scales in a later section and explain how we have arrived at the spacing of notes that we now use, but I will also point out that virtually every scale system has problems for modern music, and none are totally satisfactory for all types of music. Books can only describe, but many web sites both detail and play examples of different scales. Examples are numerous, and some sites offer sound demonstrations and details with up to around 100 examples of scales used across the world.

CHAPTER 2

MUSICAL DEVELOPMENT ASSISTED BY TECHNOLOGY

Western Religious Music in the First Millennium

Historically, in Western religious usage of music, the earliest docu-mented examples are for synagogues of a cantor (singing in a limited range of styles) together with a limited number of instruments, and a small male choir. The pattern has existed since around 1000 BC. It was not totally rigid, as improvisation was allowed.

Christian church music took a rather different, and simpler approach, both in churches and monasteries. Over the first thousand years not much happened. There would have been new hymns composed, but the singing was very simple, either with one person, or perhaps several monks singing the same line together. Church buildings were often large and echoic, so the music was slow, so this singing in unison was accept-able. It was called plainsong. Slow music would have seemed appropriate to the content of the psalms, hymns, and so on. Since there were monas-tic services held throughout the day and night, one assumes that the repetition of such events included both Te Deum and tedium. To minimize the somewhat boring aspects of these events the more adventurous musical monks then introduced the use of two notes being sung together in parallel. The options were to have the two parts exactly one octave apart or separated by a musical fifth. Although initially resisted, such moves evolved into more variations, where the two parts did not neces-sarily sing identical lines.

There were many innovators in these small changes, but their names have mostly been forgotten (one consequence of only singing from memory). Exceptions were taking place in the 12th century. Léonin added not just a second part singing the same words in parallel, but also used some variations and/or different words. Léonin may well have been the Parisian poet called Leonius. The French for *word* is *mot*, so the term *motet* emerged. One reason for parallel singing, separated by an octave or a fifth, is that male voices differ greatly, and particularly when there are young novices, there is a big range of available voices. The next step was the addition of several notes as a musical backing. These we now call chords. This was the switch to add some harmony.

In a mere thousand years the church music had moved from having a single line sung by all the monks in unison to having some duplication and/or chords. This was the onset of a style called polyphony. Polyphony initially meant that two or more parts were either sung at a fixed interval, of perhaps an octave or a fifth, or for the truly *avant garde* monks, different singers had identifiably different parts. Here the pace quickened, and over the next 500 years polyphony became ever more complex, and it developed very strict rules on how the music could differ between the parts, and also how many parts could be used.

One should also mention other key players in the development, such as the German Benedictine nun, Hildegard of Bingen, who wrote in the early 12th century and expanded away from plainsong into more melodic and tuneful works. To us this seems trivial, but at the time it was a phenomenal break from staying with the same chants from many earlier centuries. She also used clusters of notes (i.e. chords) to underpin the solo voice line.

By the 13th to 14th century onwards the types of chord became less restricted, and an early innovator was Pérotin (around 1200 AD). He benefitted from hearing music in Paris brought there by the many travellers from distant regions of Europe and Asia. By the 14th century another Frenchman, Guillaume Machaut, had expanded motet writing up to four parts (i.e. four lines of song in parallel). This example of multipart polyphony steadily went out of control. An extreme example is for a 40-part motet, 'Spem in Alium', written by Thomas Tallis around 1570. It is still performed today as a *tour de force* of choir singing. It may be fun to sing, but I am clearly missing something, because to me it always sounds a confused mess of sound, and the individual voices just blur into a background noise. Nevertheless, the skill in writing is impressive. Polyphony survived up to the time of Bach, and he was probably the high point of the discipline. The complexity of the rules, and the skill needed to write good polyphony sowed the seeds of its destruction, and post Bach it went out of fashion.

From our viewpoint it is clear that polyphony was only feasible because there was a clear musical notation; such a style of composition could not survive solely via human memory. Technological innovation was already a crucial element in musical development.

Musical Notation and Printing

Some 2000 odd years after the Chinese had named pentatonic notes, there were attempts to find a musical notation, that allowed either religious or secular works to be played without having had direct contact with a previous performance. Various attempts were made, both in Europe and India, that had partial success. One really crucial step forward was introduced by the monk Guido d'Arezzo (995–1050). His notation improved on earlier trials and, although it may look difficult to understand for a modern musician, it nevertheless was a workable attempt to portray the relatively simple church music that was in fashion at that time. One reason that he made an immediate impact is that he claimed that his system allowed a singer to learn the church repertoire within two years, rather than the ten that was then the norm.

At the time of Guido d'Arezzo printing was non-existent and it required a careful scribe to copy the music for distribution. The skill needed to make the paper, grind oak galls and so on, to make a black ink, and sharpen quill pens was clearly very tedious, but presumably good scribes had assistants to help with such support tasks. From our perspective we may imagine the processes were very slow, as well as needing great care. However, we probably underestimate the efficiency of these scribes. An exceptional example is the Doomsday book, which appeared in 1086. It listed the inhabitants, properties, and possessions of the Anglo-Saxons who had been conquered by the Normans. This information was written in two books totalling around a million words. The writing was completed within the timescale of one year. This is not just impressive, but it is in the handwriting of just one man. We therefore may need to revise our views on 'mass production' of texts and music if such feats were routine.

Guido d'Arezzo was clearly an innovative man, and he wrote training music for singers. In particular, he produced a hymn to St John in which he successively raised each line by one note. The first syllable of the line

Table 2.1 A hymn to St John

UT *queant laxis*	That servants we
RE*sonare fibris*	with loosened voice
MI*ra gestorum*	miracles and power
FA*muli tuorum*	of thy deeds may praise,
SO*lve reatum*	take heavy guilt
LA*bii reatum*	from defiled tongues,
SA*ncte Joannes*	Saint John.

was then used as the name for the note. The text he used (Table 2.1), and an approximate translation by Galston, are as shown in Table 2.1.

UT did not flow easily, so this was replaced with DO when singing, but it is retained in naming key signatures. We still recognize this labelling today of his naming of the scale notes; we use the term sol-fa. Personally, I prefer the mnemonic sung by Julie Andrews in 'The Sound of Music', but any system which has survived a millennium is impressive.

A Letter to the Corinthians

In early church music the singing was primarily from the clergy and the choir, partly because it would have been in Latin, which was not understood by the masses, and so added to the mystique. Choirs were not mixed, and women did not sing in church services (except in nunneries). To a modern churchgoer, it may be odd that women were not even allowed to sing in church, as for example, the current Catholic Church is roughly 60–75 per cent female. Their rejection, and all their consequent exclusion from any important church decisions seems to stem from just a few verses written by Saul of Tarsus (St Paul), in his first epistle to the Corinthian church (chapter 14 verse 34). The translation is approximately that women should not be allowed to speak in church, and by implication have any say in the organization of it, as this was totally the prerogative of men. My 21st century view of St Paul is that he was an insecure misogynist, but a kinder opinion might be that he was merely reflecting the current philosophy of both Greeks (Corinthians) and Turks (Tarsus) from the previous few centuries. Nevertheless, the problems he caused for women, and society, have run for the next 2000 years, and are still only too evident.

As church music developed, the sounds ranged from the lower registers of male monks to higher pitch singing of boys who were trainee novices. The church music inevitably exploited the differences, but with a slowly learnt skill in singing there was a shortage of high voice boys. The obvious solution, as in popular and folk music would have been to have female singers. But this was excluded because of the edict of Paul. The church solution was dramatic, and it caused a great deal of pain and suffering for many men, and had a significant influence on music, solely because the church wanted music with high notes. Further, many composers (including Handel) wrote operas where not only female parts were sung by men, but also many men's roles were sung in very high registers. The religious route to produce these upper notes that sounded so well in big church buildings was to have castrati as the singers. The practice of producing castrati for church music survived (albeit technically illegally) in the Vatican up to the 20th century and the last such man, Allessandro Moreschi died in 1922. In the 18th century some 4000 boys per year were castrated in Italy, in order to achieve a high vocal line with the power of an adult.

Apparently, the timing of the operation was difficult to assess. Too early into puberty and the voice just stayed as a weak treble, but later on it could have considerable power. For the successful, such as Farinelli, the reward was considerable wealth and musical fame. Less obvious is that with good 'surgical' timing the castrati grew quite tall, and functioned well as a male (except they were sterile). This apparently gave them a highly active social life.

Polyphony in Secular Music

There was a steady progression to complicated situations where different singers were singing independent parts in all types of music, not just motets. The church was not the only source of music since there were folk songs and travelling singers, the troubadours, who might have sung to their accompaniment on a lute. These minstrels travelled around Europe and also further afield, for example to the Crusades or to hear the Arabic music via Spain. Their influence was to disperse a wider range of songs and styles. The fragmentation of the church with the rise of

Protestantism also had an impact as services were translated into the local language, initially German, and then English and others. Hymns were written with different emphasis (e.g. even by Luther), and were sung by all members of the congregation, including women. Musically new harmonies had emerged using intervals of thirds, and there were conscious differences between major and minor scales, and groups of three notes (triads). Overall, music had progressed by the end of the 16th century to the point that it had strong resonance with our present day musical tastes.

While the top composers of their day may have faded into obscurity, or only be remembered by a very few people, there are music groups who continue to play or sing the motets and madrigals of that period, and a few pieces may be recognizable by a wider public. For example, the tune Greensleeves, which was incorrectly attributed to King Henry VIII, is instantly recognizable some 500 years later. Similarly, if we hear music of that period, we can probably guess the vintage, even if we do not know the composer. In fact, this is probably true of music from most periods.

Henry VIII may not have written Greensleeves, but he was indeed very enthusiastic about music, and is reported to have had a vast collection of instruments, comprising some 272 (wind) and 109 (string) examples.

More Recent Developments

In most respects, one of the greatest composers with polyphonic training was J.S. Bach. He may have been the ultimate peak of this genre, but he is equally at the end of such a compositional style. Despite this, modern musicians perform his work, and that of his contemporaries, such as Handel. Similarities between Bach and Handel include the facts that they were almost identical in time span, and both went blind in old age, and were unsuccessfully treated by the same physician. Nevertheless, they never met and were rather different in musical character. Both were superb performers, and able to improvise extensively. In the earlier part of his life Bach only focused on church music (it paid his salary) and he composed with the complex musical rules for counterpoint and fugues. These are works which seem to have mathematical transformations as they modulate from one key to another, the themes progress between instruments or run in parallel, but displaced, and there are inversions of the tunes. Some sections could even be played with the music upside down, or backwards, and yet have many of these characteristics. (In fact, if one does this, the music is still clearly by Bach!) Realistically the complexity of polyphony and fugue, with the highly developed rules or guidelines for such compositions, limited the appreciative audience to those with musical training. Such music had sincere religious integrity which was commensurate with the period, but much of it was not music that had mass appeal. A greater emphasis on tunes was emerging, not only in folk music, but also with the birth of operatic style works (as with items by Handel). There was then the start of a more familiar style of more symphonic music, not least by the next generation of Bachs. These included two of his sons, Carl Philipp and Johann Christian Bach, who both worked in London.

Singing, Secular Music, and the Birth of Opera

In addition to all the church-funded musical developments we need to remember that there were musical developments taking place in secular music. This spanned music across all social classes and focused on aspects as diverse as love, the countryside, and life and death. Except for troubadours, and people travelling as a result of warfare or shipping, it was generated and retained by a very small and static population. Consequently, much of the material did not survive, or was alien to other regional groups, even within the same country, but with different dialects. Instrumental accompaniments were limited to simple items, such as lutes, and these had low power, so were unsuited for anything except very small intimate performances. This situation changed as part of the Renaissance liberation from church domination. There was a greater focus on individuals, and the development of operatic style music. These new formats drew inspiration from a wide range of topics in art and literature. In particular, there was a tendency to look back to earlier generations such as the mythology or history of the Greek and Roman times, as well as their theatre. Part of this development meant that there was a mixture of words, plays, and music within a single performance. Many of the events were not constrained by a single theme during an entertainment, and the introduction of ballet was often an added feature. The writers, sponsors, and audience would have been from the wealthier classes with 'classical' educations, so this favoured the Greek/Roman topics for the productions.

Consequently, many stories with music were based on the Greek tragedies, and the production of Dafne, by Peri in Florence in 1594 may have been the first true example of something we would recognize as an opera. Peri later collaborated with Caccini on a version of Orpheus in 1600 and by then opera had arrived. The centre stage moved from Florence to Venice, where Monteverdi was a key and successful proponent of the early opera format, with versions that are still being performed today. Italy was totally taken by opera, and major cities all had several opera houses. For cities such as Venice or Naples, there could be as many as eight different opera performances per night (N.B. that is in a city with a tiny population of ~125,000 in 1600). With such a proliferation of demand, there was a corresponding mass production of compositions. It has been claimed some 40,000 Italian operas were written and, not surprisingly, some were so bad that they fortunately vanished extremely rapidly.

Survival of Fame and Compositions

Not only have poor quality operas faded from the musical repertoire, plus folk music that was never written down, but a similar fate has befallen the majority of all styles of ancient music. Some music, like that of J.S. Bach, was at the climax and end of an era. He has partly survived as he was involved with the new technology of keyboard instruments, and he made a major contribution to subsequent keyboard playing. He is famous for taking advantage of, and advertising, the fact that keyboard instruments were being played with our modern 12 semitones within an octave. Most importantly, the tuning was being adjusted (i.e. distorted) so that music could modulate between different keys. Bach wrote a definitive set of works in all possible keys, which has helped to entrench this keyboard tuning into our musical literature (his famous 48 Preludes and Fugues for the Well-Tempered Clavier).

This new tuning, called 'equal temperament', is so basic to modern music that I will need to examine it later in more detail, and show how and why it differs from the idealized 'pure' scales of Pythagoras, or indeed from scales that are routinely used when singing or playing string instruments if they do not have the constraints of a fixed keyboard tuning. Despite the universal use of equal temperament, we need to recognize that modern piano tuning does not totally follow the concept exactly, and there are many further compromises and active mis-tunings which are probably unknown to the average listener.

Bach's music in terms of performance and contrapuntal composition was excellent, but he did not venture into the operatic style music, which was just emerging, and his works were often too complex to be appreciated by the general public of his day (or even now). He spent many of his later years assembling and packaging his compositions, not just for commercial sales (where he was successful), but with a clear view to posterity. Despite this foresight, his music dropped almost totally out of fashion, and it only re-entered our repertoire as the result of efforts in the early 19th century by people such as Mendelssohn.

Handel was similarly a skilled performer but had a very entrepreneurial approach to his writing. Opera had emerged as an art form (particularly in a style from Italy), and this meant stage presentations and a story line that was not constrained by religious topics. It also meant that both men and women could act and sing together. Handel was financially successful, but rather than rely on church funding, and all the political problems that were encountered by Bach, he had continuous royal patronage once he moved to London. This included four successive monarchs. Handel had an advantage in obtaining royal patronage compared with the local Englishmen, because it is claimed that George I initially did not speak English. At first, Handel was commercially secure with writing of music in the Italian style of opera. Unfortunately, there was a sudden swing in public interest away from these operas, and a more puritanical attitude to women singing, so this format (and money) suddenly declined. Handel had business acumen, and switched to oratorios. These were easier to stage and, since they had biblical themes, there could be no criticism of having female singers. Handel coped well with changes in fashion, whereas other composers did not. His operas can now appear very dated for a modern audience, not least because he wrote many male parts for counter-tenors (or castrati), whereas we now expect the lead roles of our operatic male heroes to be in much lower registers. Lower voiced males are apparently considered more authoritative by men and have more sex appeal to women.

It is worth noting that he was working at a time when the general public was attending musical events and paying to do so. This opened the door to music which was neither religious, nor aimed only at those with education in Greek or Latin. There was thus an opening for a totally different style of commercial music. One populist example was 'The Beggar's Opera' by John Gay. The plot was recognizable life to the mass market.

While Bach had a score of children, and some continued the family tradition of being high grade composers and performers, Handel did not leave such a following. He did however fund many charities related to children.

Sudden changes in fortune of composers and performers were, and are, among the major disadvantages of not having a lifetime patron (as for Haydn), or church funding. The problem was epitomized by the changing fortunes of the composer and violinist Corelli. He went from being a top star composer to a minor one within a very short time, as the public were demanding more virtuosic compositions and performances. His music was initially considered difficult by the then current standards, but the number of able performers rapidly increased and violin techniques advanced. He is also credited with being one of the first to have a small orchestra in a relatively modern format.

The fortunes of Corelli were varied. He was one of the first to be able to use type printed music, and so increase the marketing and sales of his music. He introduced the use of a shorthand called a 'figured bass'. This notation meant that in accompanying music, especially with church organs, the pattern of the bass chords was well typified and understood. So instead of writing out the entire chords he had a simple notation of a letter (for the main note of the chord), plus some number annotations such as 6 or 7 to say one had to include not just the basic home chord of say 1, 3, 5, 8 (e.g. C,E,G,C), but add in the 6th or 7th to add more tone colour (A or B). This was a period when any skilled performer was expected to be able to improvise with a very high level of musicality.

Figured bass died as most musicians stopped improvisation, but one can see that it became reinvented in the 20th century in mass market music where the manuscripts may well lay out all the notes and text for a song, but the accompanying guitarist just has a coded message giving the same instructions as for a figured bass.

A clear example of the speed of change in taste and performance is that the violin music of Corelli and his contemporaries did not venture to very high notes, or use the upper parts of the violin strings. (For violinists, the description is that he rarely went beyond the third position, i.e. an extremely modest challenge in 21st century playing.) Apparently, when presented with a work that explored the upper registers of the violin, he was not able to perform it, and in a very short period of time he was out of fashion.

Many other composers who were equally famous within their generation have almost totally slipped out of popularity and memory. Buxtehude was such a famous organist that Bach, and a friend, took leave to walk 250 miles (each way!) to hear him perform and learn from him. Today, some Buxtehude organ works may be played, but outside of the world of organists he has virtually vanished. Another extremely prolific composer of that period was Georg Telemann (1681–1767). He was contemporary with Bach, a friend and godfather to some of Bach's children. At the time, Telemann had a far higher reputation in Germany. His portfolio of compositions is probably the largest on record, with some 1000 instrumental works, 120 concertos, oratorios, and operas, as well as vast numbers of works for smaller groups and individuals. His success may have been linked to his ability to write for the increasing musically literate middle classes, as well as for the public performances. Nevertheless, the pattern is the same, and within 30 years of his death he was virtually forgotten. Since then there has been some minor recent revival of his work. I personally like his music and, as a violinist, enjoy his violin concerti.

The need for patrons or church funding did not apply to Albinoni (a contemporary of Corelli), as his family were wealthy Venetians. But this posed a different problem, as he wished to be a performer and, in Venice, this was only allowed for members of the Musician's Guild. Presumably his class status blocked him from this, and so he turned to teaching and composing. He generated a very large musical output, and much of this eventually found its way to the Dresden State Library. Unfortunately, the building was destroyed in the bombing during World War II. One of the few scraps of music that was salvaged included part of an Adagio, and this is among the few pieces that he is now remembered for.

While in Venice one should note the contributions of Antonio Vivaldi, although nominally a priest (famed for his red hair) he spent most of his life teaching music to young girls. His violin playing was very advanced relative to his contemporaries, and he made great efforts to duplicate familiar sounds (e.g. birds, thunder, and storms), and set out to portray musical backgrounds of storms, winter and so on. He is particularly remembered for one such a tonal description in the Four Seasons, and rightly so, as this was a totally new excursion into the way music could be descriptive. He also wrote a vast number of violin concertos. A small number of manuscripts of the violin concertos still exist (80 were in Dresden). His critics have unkindly said he wrote the same concerto 400 times; this criticism can be addressed to most writers who work closely in the same genre on a mass production scale, whether they be music or popular novels. Less well known is that he also composed 38 operas (he was in Venice), but these have faded into total obscurity. His experimentation in music production exploited the vast cathedral of St Mark, which is extremely echoic, as the sound bounces around. He wrote items with segmented orchestral groups, in different parts of the cathedral, to give effects that are not possible in normal concert hall venues.

A far less familiar name to us is that of Johann Vanhal (1739–1813), a Viennese contemporary of Mozart and Beethoven. He did not have patronage but was financially successful and highly rated in his time. His output was considerable, not only with music for home consumption, but also some 79 symphonies and such-like! Despite this immense output most of us cannot quote a single example of his work, even if we have heard of him.

I have picked these examples to emphasize that musical quality and output are probably less important for the long-term influence of a composer's music than we would at first assume. The examples I am using here are for composers who span the period when funding of professional musicians was moving from being controlled by the church and wealthy classes, with private orchestras and other groups, to income from performances that were aimed at the general public. This meant that the style of music had to appeal to a wide audience as they were buying the concert seats. It also was a period where printed music for home consumption dropped in price to a level where a wider public was able, and willing, to afford it. One should not forget that, except in a few major cities, public performances of substantial works did not exist. There were no broadcasts or recording systems. Further, simple music, of only a modest difficulty, had to be suitable for home performance by an emerging middle class, and this was a lucrative market.

Payment for composition is highly desirable, but later views recognize that if someone is truly dedicated to a profession, whether music, art, literature, or science, then they will produce even if the remuneration is low. This is a facet of human nature which is often exploited.

What is far less obvious to us now is to guess at the quality of the playing, either by the home amateurs or in the public renditions for these earlier generations. The small orchestras available to Bach or Haydn were unusual in that they were professionals with experience in playing together. One assumes that they were competent, but these groups were the exceptions rather than the rule. For the majority of the musical events, many orchestras assembled from a relatively random collection of players who came together for special occasions. In terms of rehearsals there were few, or none! Players were often opposed to rehearsals as they were not necessarily paid for them. A single rehearsal might have been the most that was feasible. An additional feature which we might not have considered is that the music, especially for new compositions, was rapidly hand-written on manuscripts (together with ink marks from the early style pens). Extreme cases of last minute writing are well documented for both Rossini and Mozart, who are known to have been composing on the day of the performance. Beethoven sometimes performed the piano part of his violin sonatas at the point where he had only just finished writing the violin part and had not yet written the manuscript version of the piano part. The concerts were then played at sight(!) in the evenings by candlelight and read with eyesight that did not benefit from modern spectacles or contact lenses. The potential for misreading the notes was immense. Today, I doubt that any high-grade orchestras or conductors would want (or dare) to publicly sight-read a new work under such conditions.

The shortage of permanent orchestras also meant that, except for a few popular items, neither the audience nor the orchestral players would have the opportunity to hear a symphony, or major item, more than a few times in their lifetime. For a modern audience, where we have exposure to radio and CD, and can listen to repeats, and grow to like and understand a composition, the brief encounters needed something different to hold our attention. One such feature was the growth of more showmanship and more virtuosic playing. Spectacular and novel cadenzas and improvizations were then expected, whereas for modern classical concerts, spontaneous cadenzas are incredibly rare, or non-existent.

This is a pity, but we can sense the excitement of such instantaneous composition from the improvizations in jazz where the spontaneity, and the danger of success or failure, are still current practice. Overall this early focus on showmanship and spontaneous additions meant that soloists who could make an impact by such techniques were often more important than much of the music. This sold the seats. It still does.

CHAPTER 3

MUSICIANSHIP AND THE CULT OF PERSONALITY

The Growing Cult of Superstars

Part of the change in musical appreciation was that operas attracted audiences from across the social spectrum, and not just in concert halls but also in places like the London Vauxhall Gardens. Attendances were more like modern football crowds, and there were stars that were idolized and cheered. Singers who were not appreciated were booed or had vegetables etc. thrown at them. Not only had music moved from anonymous church music, but it had swung to the other extreme with a developing cult of the star performer, just as for the instrumentalists of the orchestral concerti. Superstar showmanship, and promotion, were therefore similar to the situation which is still with us today. Equally, such personality cults, whether of singer or composer, meant that fashion and publicity were often fickle, and far more critical to financial success than the quality of the music and actual performance.

Opera singers are generally just the intermediaries presenting the works of composers. This does not mean they lack star status, ego or temperament but at least for opera they rarely produce new items for posterity. In later years a few composers of operettas and musicals, such as Ivor Novello or Noel Coward, have succeeded in both roles. For modern popular music it is probably the norm that the music is written by, and/or for, the performers. One contributing reason for this is that they often have no formal training in singing, and so need music that matches their particular voice and style.

This is not the case with classical instrumentalists. They still need structured training and, as part of this schooling, many learn and develop the skills of composition. For a successful and well-funded career in the 18th/19th century they also needed the star qualities and charisma of public performance. Once again, this was more important than the music. As a pianist, Beethoven was such a man. He could improvise, or add cadenzas, that far eclipsed normal performers of his time. For the violin, there were a number of virtuosi whose technical prowess could not be demonstrated by the work of existing composers, and therefore they had to be highly proficient at writing for their own instruments, to show off their skill. Many of their concerti and showpieces have faded from the modern repertoire as musically they were not always outstanding. Even more important is that the level of technique of modern professionals has risen so much that the earlier showpieces now appear technically less impressive. For example, many of the 'standard' professional orchestral violinists etc.of today can probably play, even the solo parts, of concerti that were considered virtually impossible in the mid-19th century. We should not forget that a top violinist of his day, Joachim, asked for revisions of the Brahms violin concerto in order to make it simpler to play, whereas it now would not be rated as excessively technically difficult, and certainly not in the same league of difficulty as many recent violin concerti.

One of the most important violinists at the end of the 18th century was Giovanni Battista Viotti (1755–1824). He was taught via a sequence of pupils and teachers dating back to Corelli, and he in turn had pupils who led to Joachim. Technically, his works were more flamboyant and virtuosic than any of his predecessors, and he has been called the father of modern violin playing. His use of higher registers meant there was a need to modify existing violins to have a longer fingerboard, so that he could play higher notes. This in turn led to other structural changes, and some significant differences between violins of today and those as originally built by the top makers such as Stradivari or Amati. The surprising feature is that the instruments of Stradivari et al. have survived the structural modifications and are still incredibly desirable. The few hundred that remain are marketed at prices into the millions.

In the case of both the violin and the piano the external shape may look very similar in earlier and modern instruments. Among many changes in design, for both instruments there was a demand for more power, and this was achieved by increasing the thickness of the strings. The consequence was that a greater force is put on the piano frame, and similarly on the violin body. The piano approach was to move to steel frames, whereas the violin had to be internally strengthened. Failure to do so meant collapse of the instruments. The pianist Liszt was noted for destroying early wooden frames, as a result of his energetic playing and use of heavy strings.

Viotti played a Stradivarius violin, which still bears his name, and his fame was enhanced by performance of his own numerous concerti. However, it is hard for us to recognize and accept that the role of the violin was not as it is today. When Viotti went on a concert tour to Paris in 1782, the local press was highly critical about the prospect of a pro-gramme for violin music, as they claimed it was such an inferior sound-ing instrument compared with the viol! Viotti thought otherwise, and his performances and exhibitionist playing totally transformed the Parisian scene, so that the violin became number one of the stringed instruments. His concerti are now rarely played in concerts, and are no longer seen as technically challenging. Indeed, except among violinists, his name may be unknown.

Many other violin virtuosi have followed the pattern of Viotti, with performers such as Vieuxtemps and Paganini. Vieuxtemps, like Viotti, has faded from public view. His music can sometimes be challenging and is enjoyable to play. On the downside it often sounds dated and unfash-ionable. By contrast, Paganini was a far better showman. He cultivated an image of a tormented performer, and dressed in black to emphasize that role. The modern pop star image would be Goth. Publicity stunts included rumours that he had been taught by the Devil. This all added glamour and excitement to his appearances. During his lifetime he kept the manuscripts of his concerti out of public circulation, and this further boosted his mystique. The net effect has been that his name is far more familiar than that of Viotti, and his music is still firmly in the modern violin repertoire. My only caveat is that while Viotti concerti are straight-forward to play (e.g. as training pieces), many of the Paganini caprices for solo violin rapidly separate those who can overcome the very high first hurdle, and can play the notes, and those true virtuosi with the real skill to make them musically exciting as well. But the lesser performances are still applauded, as even the non-critical audiences recognize the sheer level of technical difficulty.

Stardom, and the fame that goes with it, is often transient, but some early superstars are remembered even if only for a single event that shows how they moved the audiences of their time. In this category I will again mention the castrato Farinelli (whose real name was Nicola Broschi). He was clearly exceptional, as it is reported that in 1737 he came to London, started an aria written by his brother with one dramatic and controlled *single* opening note, but the audience interrupted and applauded it for five minutes before he could continue! Farinelli has continued to hold a musical niche, even though we no longer have any examples of the voice tones that such castrati trained singers could produce. His continuing fame includes a film about him. Other superstar musicians have similarly had films created around their lives, such as Paganini and Mozart. The films make an unexpected social comment on the way our appreciation of music and films has changed within 50 years. Those who saw the original screenings, and then a reissue, frequently say the older films and scripts have dated very badly, but the music has not.

It is difficult to decide why some musicians have had star status while others, equally skilled, have never hit the public taste. In the 19th century the great pianists Chopin and Liszt both caught the public imagination. Their fame raised them above the normal view of musicians being mere servant class, and writings about famous musicians often treated them as some type of tormented artist. In their time, 'public' was often a more limited salon audience, and their impact was by direct close intimate contact with the audience. They both cultivated dress sense, publicity, and image, to mark them out from normal people of the day. This romanticized publicity highlighted their life styles, and put them in the same category as current film and pop stars. For my personal taste, the modern performances of Liszt often highlight technique, high speed, and volume, rather than musicality. Pianists who can add that extra finishing touch are rare. Indeed, this was a criticism of Liszt himself.

The cult of superstar personalities in music may have started to emerge after the Renaissance, and so it has existed for several centuries. It is very much a key part of all modern branches of popular music, theatre, film, and TV, but it is less obvious whether the current 21st century stars will have sufficiently distinct image and qualities to enable them to be remembered in later decades, let alone later centuries—not least because all the records and films etc. will have been superseded by formats that we cannot even imagine at this stage. Therefore, the fact that they are widely distributed on modern recordings may actually be the death knell for the long term, as once new recording formats emerge, they will probably vanish.

This is a definite downside of rapidly developing technologies, and I will mention it in more detail in a later chapter.

The Cult of Conductors

Not only singers and instrumental performers benefitted from the early 19th century public desire for musical heroes. The same is now seen for conductors. The baton and early conducting techniques emerged initially through Spohr and Beethoven. Jean-Baptiste Lully was a highly successful composer, and bandmaster at the court Louis XIV of France. In outdoor performances he used a long pole (with a spike like an Alpen climbing stick), and in 1687 he damaged his foot with it, and died. Other orchestras were conducted by the leader of the violins with a bow. Large modern orchestras tend to have a conductor with a short baton, although some prefer to wave their hands without any baton.

Some conductors specialize in particular genres such as ballet, opera, choirs, or large orchestral works. This is a little different from say Beethoven or Wagner, who mostly conducted their own music. A good conductor does not just beat time in an attempt to keep all the musicians together, but also defines the variations in dynamics and tempi, and should impart a coherent overview of a particular piece, to inspire the performers. This is either easy or difficult, depending on the quality of the orchestra, and the way he has sold them his perception of the piece. It is not a simple task, especially if the orchestra has frequently played the work with other conductors. Freshness is lost, and/or they do not readily change for the demands of a guest conductor. Conducting has expanded to include women on the podium. In the modern world many nationalities and languages are involved, and communication can be problematic. Conducting is a tricky business because success is often credited to the orchestra, but failure to the conductor. One result is that some conductors have taken an excessively dictatorial role, and run publicity campaigns, and make demands for fees that far outweigh their ability. In order to succeed, the recording companies (who are paying the fees) attempt to provide even greater hype about their superstar. Once the public reads the fantastic claims about this wonderful conductor, they buy more from the same source and ignore superior recordings from lesser name maestros. Image is the name of the game.

Many major conductors have emerged who, while excellent professionally, have had a mixture of egotism and prima donna temperaments that have added to the publicity and hype of marketing to a degree that raised their image far above that of their contemporaries. Financially, for them and their recording companies, this was a brilliant strategy, but it is far less apparent that it was good news for other conductors, or for the record buying public. For example, in the USA Toscanini was considered as 'god', and other excellent conductors were left on the side lines.

The opinion of orchestral players does not always agree with the marketing and audience view of many conductors, either because the conductors may have achieved their aims in the rehearsal, so the performance does not need a dramatic piece of stagecraft, or histrionic conducting may not closely relate to the signals needed by the orchestra.

I will separate conductors into several categories. Firstly, there were those like Beethoven or Wagner who are highly rated composers, and who limited their conducting mostly to their own work. Other composers, such as Mendelssohn and Weber, successfully conducted more general works, and also promoted items from composers who were not necessarily mainstream in their time. The less common example is Mahler who wanted to be a composer but earned his living primarily as a conductor. This probably limited and delayed his efforts at composition. I personally think his symphonies are enjoyable but they can sometimes seem over long, for both the players and the audience.

Many conductors must be congratulated, not only for their bringing lesser known works to the public arena of classical concert goers, but also for greatly widening the market to people who would not have thought they liked classical music. They have had immense influence on the range of music that we now appreciate. Many names could be listed here. People such as Stokowski in the USA and Mackerras in the UK used their compositional skills to arrange and popularize less familiar composers, and present them in an approachable form. Mackerras was appreciated by orchestral players as he annotated the scores much more than normal, so as to reduce the rehearsal time that was needed. Sir Henry Wood and Sir Thomas Beecham are similarly names of conductors who created an outstanding popular appeal of music to a mass market, not least via the annual Promenade Concerts at the Royal Albert Hall.

There is a final class of conductor which includes excellent soloists (piano, violin, singer, etc.) who have passed their soloist performance

prime but feel they understand music, and therefore can become conductors. If they were big name stars, this is good for ticket and record
sales. Unfortunately, such late conversions are not always successful.
I have heard extremely critical comments from orchestral players, including views that 'the concert went well—because we ignored him'. On
recordings, the role and physical input of the conductor is hidden in
these situations, so the sales rise because of the name. Live concerts can
tell a different story from edited recordings, and I clearly recall hearing
an eminent (albeit elderly) violinist turned conductor becoming quite
lost. The orchestra (excellent) carried on regardless and finished no more
than two or three bars ahead of him. The concert was broadcast, and
from the radio version the conductor was praised.

Feedback from Critics and the Audience

People attend concerts for many reasons. For some, concerts are social
events where they can be seen and meet friends. Others go because they
like the programme, the performers, or the composers. Probably in all
cases a key factor is initial guidance that they have read or heard from
critics or friends. The role of critics is therefore very influential in defining popular taste as it can either harm or promote the careers of the
composers and musicians. It can equally encourage the introduction of
new musical styles and works from composers who may have faded into
obscurity. A century ago the only way to receive such critical comments
was by journalism, but now broadcasting and internet comments are
readily available. Even live broadcast performances can be heard for a
second or third time.

My date choice of 'a century ago' is because that represents the
availability of gramophone and radio, which had an immense influence
on both the general public and those who were already interested in
classical music. It suddenly became possible to hear new works, and to
offer music to people who previously had neither access nor money to
attend live high-grade concerts. The 1920s were similarly the time when
electronics had entered recording, and the quality of gramophone recordings had started to improve. The original popularity may have been mostly
for dance music entertainment (the disc recording times were only a few
minutes), but the exposure to music opened new horizons. In terms of
appreciation of music this was a crucial time.

Both for new works and items that were not in the current repertoire I have already emphasized that 18/19th century performances often had minimal rehearsal, so neither conductor, orchestra, nor audience heard ideal performances. Worse is that as humans we respond to new ideas with something which psychologists describe in terms such as cognitive dissonance. All this means is that if we hear or read of something new (whether music, science, or politics etc.) we will mentally consider it and possibly accept it if it is close to our pre-existing experience and beliefs. However, any totally new idea or musical sound we automatically reject as our first response. This is as true not just in philosophy or music— where there is always a highly personal interpretation—but also in science, despite the fact that there may be hard factual evidence. In science, many novel and truly innovative ideas can take 20 years to move from rejection to being mainstream dogma. Music is no different. Reading about the reactions of critics and audiences to new works says that many were often totally condemned at their premieres. The many examples include both instrumental works and operas, even when they were in styles current at the time. Less surprisingly is a general hesitation to accept music in newer styles, particularly if it has new tonalities, or other gimmicks.

Rejection can be triggered by poor quality first performances. Carmen, La Traviata, and Rigoletto all can be listed in such first night 'disasters' for a variety of reasons. Carmen was unacceptable to the French and to the Opera Comique in Paris because it included a death on stage. This was considered to be very bad luck by the superstitious managers and performers of the time. There was also a more subtle social problem that the main protagonists (with all the good tunes) were not nobility, but soldiers and cigarette girls. The first night of Traviata struggled from poor casting of the singers, not least because the consumptive heroine, Violetta, was sung by a substantially built lady who did not fit the image of the part. The producer also altered the time setting by a century so it was both inappropriate and had lost the power of the moral comment that was essential for success. This was unfortunate as Verdi had consciously chosen a topic to highlight the hypocrisy of the morality related to courtesans such as Violetta.

In Rigoletto there was political restraint imposed, as originally the licentious villain should have been a French king, and this was considered unacceptable and unpatriotic, so there was a libretto shift in country and time to make the villain a 16th century Duke of Mantua.

All these operatic examples fortunately had better success in later offerings and are now securely in the modern repertoire and in the 'top 100' favourite works.

Audiences are humans, so the cognitive dissonance effect has always operated like this, and the list of composers who suffered from it includes current establishment favourites such as Saint Saëns, Ravel, and Stravinsky. All had compositions that were initially seen as too avant-garde, and their works were slated by the critics. Now they have slid firmly into the sound of 'classical' composers. Does this mean that the items we now think are dreadful will one day appear equally feted? Maybe this is a benefit of a finite life that we do not have to endure such events.

Trusting the journalistic critics will also involve a minefield of difficulties. Critics are not like the general public in terms of their musical knowledge, experience, and taste. It is their career, so they will hear and experience events in a manner closer to professional players than the average audience. If we compare this with languages, then the native speakers understand, and expect, faster delivery than tourists can manage. Similarly, professional musicians (and critics) have speed and tonality expectations, and a desire for innovative works, which do not match those of the general audience, who listen to concerts only spasmodically. They can therefore be poorly guided and influenced by the critics' reviews, as there is inevitably some disparity between the views of professional critics and those of their readers. Comments written by critics should be treated with caution, especially if they do not remember that their main audience is less skilled and/or more cautious in changing musical taste.

Perhaps for the same reasons, a comparison of recordings of different vintages shows that many of the familiar classical works are now played faster than they were 50 years ago. Similarly, the approach to baroque music has moved from a slow reverential pace to now revealing that much of the work was intended as speedy and enjoyable popular music, often intended for dances.

We in turn should recognize that, for their continued employment, the reviewers need to have an impact with their criticisms. We might expect bland platitudes when reviewing amateur performances, but for professional items the good critic will often write a scathing, or adoring, phrase to keep our interest. Many do this by being extreme and cynical, as this is a good career move for them to stand out from the other critics.

The comments can be entertaining but not necessarily a realistic assessment of a performance.

By contrast, critics who present and compare various versions of recordings on radio or TV programmes are in a different situation, because we are then able to directly hear the music and consider if we agree with their views. They often review many alternative recordings and suggest which they think might be ideal for our library of CDs. We hear the snippets they offer and so we always benefit from their insights and are able to consider if we agree with them. In this situation we may prefer a certain critic if we consistently like their views.

The other factor, which is purely 21st century, is that members of the general public are totally uninhibited in making on-line comments and blogs about particular works and performers. We need extreme caution in reading these on-line comments. We will rarely see enough from the same source to know if we can trust them. Blogs are often uninhibited and consciously destructive. Further, it is not difficult to manipulate the number of responses (both positive and negative) so that a specific performer appears to have a greater following (or number of detractors) than we might have expected. For recording sales, these aspects of on-line boost or attack are effective. The situation is a simple parallel to the role of the paid claque who did (and indeed still do) cheer the opera stars that they have been paid to support and/or make noise or hiss the rivals or soloists who have stopped paying them. The opera claques and the covert website fan clubs, as well as marketing ploys, all distort the true assessment of musical quality. Exactly the same distortions and high public exposure arise via TV chat shows, phone-in programmes, and email music requests. We will probably never know just how much they skew our opinions. Many radio programmes broadcast a very limited range of musical five-minute fragments, and then run surveys of their listeners as to what music is in their top 100 items. The listeners never hear anything else, and so there is a self-sustaining and very blinkered view of the music and performers that are available. It needs a courageous presenter, or enough phone-in suggestions, to break this mould. In parallel with this limitation, repeated exposure to just fragments of highlights steadily erodes appreciation for the music that is being beamed at us, as we lose their context. The same erosion of careful listening can equally apply to complete works that are repeated too often.

The bottom line is that new music, even from familiar composers, takes time to be accepted. Music in new styles is almost automatically rejected at first hearing, especially if it is very different from earlier items, but subsequent performances, and reassessment, can make even the new items seem familiar and acceptable. There is a 20-year time lag in accepting novel ideas in science, and I suspect this is the same in music. Perhaps it reflects that within 20 years the audience for both is of a different generation.

Reasons Why We Can Never Repeat the Musical Past

As we grow older, even on a day-to-day basis, we have no problem accepting that we can talk about earlier events, look at photographs, and pictures or letters, but never fully re-experience the same events in our memory exactly as they seemed to happen. Many details of names and events, even of the closest of friends, will fade or become altered. This is human memory. It may variously be irritating, or desirable, depending on whether we want to recall a happy or sad piece of history. It also means that we are less aware in old age of our degradations of the skills and fitness that we had in our youth. Slightly less obvious is that, with hindsight and subsequent experiences, we may interpret the historic details in a different light from our original view of them. This is a little philosophical, but we need to understand that this is precisely what happens to the way we appreciate music. Earlier on, I categorically said we could neither hear nor appreciate earlier compositions (Beethoven was my example) in the way that he wrote them, or in the way audiences of his day heard them. This has nothing to do with the fact that earlier music can still be emotionally powerful and enjoyable.

I want to explore this theme in a little more detail and try to recognize the factors that make history unrepeatable. It is certainly not a new thought because, at the time of Beethoven, music critics and commentators were saying that they no longer understood the music of Bach, or that the operas of the previous century were unexciting. Bach probably looked back and found he was equally distanced from music of the Middle Ages.

Tempi

I will start with a topic which, at first sight, seems unlikely to have changed with time, as this may establish which factors influence the evolution of our musical appreciation. First, I will discuss metronome and tempi markings that define the speeds for the sections of the performance. After all, every movement and section will have speed guides in the usual style of *andante* (slow) or *vivace* (lively) etc. There is a consistent pattern from say *largo* (very slow) to *presto* (very fast), so this is a start. It is not perfect, so there will usually be qualifying terms to imply not too fast or with plenty of fire. This offers some leeway to suit the taste of the performer or conductor, and the room acoustics where it is being played, which will never duplicate those of concert performance. In addition to speed, guideline words convey additional information to hint if the composer wanted a piece that is a smoothly flowing legato, or a spikier punctuated delivery. Even more precise is that many composers add metronome speeds. Assuming that the metronome of Beethoven was as accurate as a modern one, this surely should put us exactly on target for the speed of delivery? The reality is clearly different if we listen to CDs or vinyl discs and compare performances. This instantly tells us that the original metronome markings are largely ignored. Further, the conductors and performers who stick rigorously to the original metronome speeds are not guaranteed to produce the best results. Variations can be quite dramatic. For example, in a long work such as the Beethoven 9th symphony, different conductors will be relatively faster or slower in each of the movements, even if overall their timings are similar. Within my own music collection, I have examples of the 9th symphony ranging from a total time of 66 to more than 75 minutes. The 66 minute version is fast, but I know there is a version conducted by Richard Strauss that takes just 45 minutes. How he did this is not obvious. Perhaps he made some cuts, as musically it does not seem feasible in 45 minutes. Pianists' recordings of the Beethoven Moonlight piano sonata can also differ over the duration by 30 per cent.

Yet more extreme examples include recordings of the 'Four last songs' by Richard Strauss. In a comparative study of the many recordings, the total times depended both on the conductor and the singer (i.e. the same conductor differed between singers), and there was no pattern that all four songs were equally slow or fast. The extremes for 'Frühling' go from a rapid 2 min 40 secs from Furtwängler to a lengthy 4 min 5 sec from Karajan (i.e. 50 per cent slower).

Our pace of life may well be faster than in the 18/19th centuries, but musically high speed definitely is not guaranteed to be a winning strategy, as often the slower items benefit, as one can luxuriate in the subtleties of tone and rhythm and, as with speech, the delivery and excitement is not lost in one continuous high-speed delivery.

Nevertheless, we tend to associate high speed with good technique and excitement. So we assume that fast deliveries mean the best performers. (In athletics the fastest runners are obviously the best.) We may also get enjoyment purely from an exhilaratingly fast delivery. Many musicians will attempt speed as part of their showmanship for a simplistic audience. The second factor, to my mind rarely discussed, is that highly trained professional musicians can play, think, and appreciate music at high speed. For them, their musical pleasure may indeed be equally tuned to this racing circuit delivery. For the bulk of listeners, even those who are accustomed to and appreciate and enjoy music, they are not racing circuit quality, and at best are just motorway drivers. This means they lack the sensitivity and musical reaction times needed to follow the subtleties of the high speed virtuosi. The result is discrepancy in intent and appreciation of the composer, orchestra, and average listener. We are not Beethoven.

Exactly the same problem arises in the theatre for a 21st century English speaker listening to Shakespeare, as delivered by a top Shakespearean actor. The plays are long, and a tradition has developed that the delivery is crisp and extremely fast (maybe because this shortens the overall performance). With a 500-year-old script it means that most of us are hearing it as for a semi-foreign language, and even if we follow the broad outline of the play we are missing the details, jokes, and other references that are heard by the experts. There is therefore no way we can consider that we are hearing the play in the same way that it would have sounded to the Elizabethan audience, nor would they have missed the topical and historical references that were written into it. In operetta, the music and comments of Gilbert and Sullivan were highly topical. Only experts in Victorian history and politics will understand many of the comments.

Music and speech are so closely linked that it is essential to draw parallels between them during a performance. Exciting speech has rhythms, dynamics, speed changes, and silences. It is never a steady metronomic delivery. Good conductors understand this, and they are willing and able to use reduced and variable speeds, and introduce pauses and pace as dictated by the music. Some of my favourite orchestral conductors are also skilled with opera, and I suspect their sensitivity to the nuances of a text enhances the flexibility needed for their orchestral work.

The conclusion is that there are two types of audience: professional musicians who can cope with fast deliveries and ideas of a professional composer, and non-professional concert-goers who need a different set of tempi to fully participate and enjoy the music. A future compromise may exist as with modern electronics, as it is feasible to manufacture CD type electronics that will allow speed adjustment, without changing the pitch. This will satisfy both types of home audience.

Musical Training

Here is a topic where we expect problems in understanding the music in the same way as the composer or the original vintage audience. Professional composers will have undergone extensive training in the writing techniques available at the time of composition, plus performance skills with some of the instruments. They may also be actively seeking new ways of expressing themselves. Their musical background will include items they have read, played, or heard and this will vary greatly with their location. For our model example (Beethoven) it will include any music that he could have heard in Vienna, or during travelling (i.e. in early 19th century middle Europe). The same is true for his audience, although only the wealthier sections of the audiences would have travelled any great distance to hear foreign music. Composition involves both inspiration and perspiration, and some of the inspiration comes from the sounds one hears every day. These are not just music. Beethoven and his audience lived in a small city with direct access to the countryside, so everyday life included the sounds of birds and animals. These appear directly in works of the period, from examples of a cuckoo, or the many birds and sounds detailed in the Vivaldi Four Seasons. In a modern audience, from say London or New York, many people may never have heard such sounds in daily life. Instead, they experience high levels of mechanical street noise, incessant background 'music' in elevators and shops, TV programmes, and mobile phones.

A nice example of this change in background is offered by the Schubert song, 'The shepherd on the rock'. Here the young shepherd lad is on a mountain-top singing to his girlfriend in the valley below. It was a realistic possibility at the time of Schubert, but now virtually impossible with the background noises of a modern landscape.

Musical training will generally involve attendance at a music school or conservatoire, and instruction from the teachers of the period. Whether willingly or not, the students will absorb many of their ideas and stylistic patterns. Therefore, a composer or performer of a period is imprinted with the style of the period. This will partly be subconscious, but it is also encouraged if there are prizes, competitions, or exams to be passed that lead to career opportunities. The positive aspect is that the young composer will gain experience in the style which is fashionable, and therefore marketable.

I see identical behaviour in the training of scientists. The best students seek the best teachers and establishments. But in doing so, they narrow their perspectives to the current hot topics of the time. This means that they continue their working life with the majority of them deviating little from their training. I have seen studies which say that, of professional scientists working in universities, more than 90 per cent never move far from the topic they worked on during their doctoral thesis. In my scientific experience, this is a realistic comment, and I doubt that it is too different in other careers. I have worked in a wider range of topics than many scientists, but this is still only a tiny fraction of the possible topics within physics, and an even smaller percentage of areas one would describe as science. Outside of my specialist area I am invariably quite ignorant.

For scientists, as for musicians, if we read older articles and books, from say 50 or 100 years ago, there are considerable difficulties in understanding. Later developments mean that some ideas and theories have not survived the test of time, and other topics will have advanced so much that the old books will seem confused, and/or the style of writing can be disconcerting. The older texts will use words and cite equipment that we will not always recognize, or they may be used in different ways in the present day. Just sometimes there are hidden gems with ideas that were so far ahead of their time that they were forgotten or could not be developed. These may be reinvented 20 or 50 years later. Therefore, many of us have looked at early scientific works, actively seeking inspiration from these missed opportunities. Musical equivalents are the pieces hidden in archives that can have resonance with a modern audience. Just as the scientific literature (even in a single subject such as physics or chemistry) has expanded so much that there is no possibility that one individual can appreciate more than a minute fraction of it, in music, no person can have anything more than a similarly narrow perspective of musical styles and compositions. This is true even if we limit our view to just western classical music. If this seems unlikely, then consider how many composers have produced works for keyboard instruments. My challenge to a reader is to research (or estimate or guess) the number from each composer, and take an average performance time. For a 40-hour listening week, we will arrive at numbers implying that it would take decades to sequentially hear all of it. I am always amazed at how many gems appear, from composers that I had never heard of.

An interesting example of an idea that died, and then was resuscitated 70 years later, is that Liszt experimented with music that used all 12 notes of the octave before repeating any of them. The reinvention of the idea was made in the 20th century by Stockhausen.

Changing Styles of Performance

While we can only guess at the performance styles of the early music, there are many comments on the delivery of the superstar virtuosi, and critical comments and praise, or derision, about many operatic performances. Guessing from a text is hard, but with the advent of recording techniques we have a slightly better indication of what was fashionable, even when the quality of the recordings was poor. Among the earliest examples are those of Adelina Patti, a great soprano of the 19th century. At 62, she made a recording in 1905 which reveals not only that she had an excellent clean singing voice but more important than the recording quality is that it displays just how variable were the speeds, additions of embellishments and cadenzas on the music of that century. There appears to be a total lack of inhibition, both from her and from other earlier recording artists as to constraints set by the music and markings of the composer, and the freedom to improvise around it in performance. Variations in interpretation between the soloist and accompanist, or several soloists, are considerable as judged by modern playing.

For instrumentalists, there was the same flexibility in timing, many examples of rubato (i.e. a local slowing down for extra emphasis or pathos), often an aversion to, or playing of, a different style of vibrato (controlled small changes in pitch—wobbling if badly done!), and sliding notes (glissando or portamento), which we would currently totally reject. Little of this emerges from the written reviews, as none of it was unusual to audiences of that time. For us, or anyone who wishes to recreate an earlier playing style, the early recordings act as a warning that duplicating early performance styles has aspects we cannot easily appreciate, and many that would be unacceptable to a modern audience.

Who are musicians?

Clearly a trick question, as the answer can either be nearly everyone, or a tiny minority. If I define music as enjoyment of memorable tunes and rhythms that we can recall and sing, then nearly everyone has musical skill. If I limit the definition to people with formal training, or a particular genre, it will drop the numbers into a wide range of different categories. My definition of musicians does not automatically include everyone who considers themself to be a leading exponent. I particularly liked a quotation I read which said, 'The old guard of the avant-garde, are deeply suspicious of any significant move towards tonality, any hint of pulse that is discernible, and any music which communicates successfully with a non-specialist audience'. It may be cynical, but unfortunately there seems to be quite a lot of recent 'music' which fits this pattern, so maybe I am missing some subtle point. Such music, I view in the same way as art created by throwing a paint pot at the wall, or rolling on a canvas on the floor.

This may seem facetious but the quote emphasizes that our tastes in music are highly variable. Even for one individual, views are dynamically changing as we hear new works and music from different continents or with new interpretations. Once it is changed we can never revert to the earlier condition, so our perspective of music is based on today, and we can never duplicate any earlier age, even if it were in our own lifetime.

I am actively offering parallels with music and science, and for science there is an equally vast spread of opinion of what is important science, or which science is fashionable and is bringing in big research grants and/ or publicity, Nobel prizes, or TV fame etc. To some, the most important science may be in fundamental and particle physics; others will claim it is astronomy; whereas such areas may equally be described as merely being dilettante science with minimal contributions to solving the real problems of the world, and therefore a waste of money, good brains, and resources. There is truth in each view, so maybe I am also being over prejudiced against minimalist cacophony.

The main question of this section is to consider who become the professional performers and composers. This is very different from considering musicians who play instruments or listen for pleasure.

Historically, musicians were mainly funded by different patrons, and to enter a career in music was definitely equivalent to becoming a servant. The literature and biographies of famous musicians frequently include the problems this caused, either in the restrictions and demands of the employers, or the difficulties in terminating a contract to work in a different city. It was equally unfortunate that for the more leisured classes, where the children had access to instruments and good tuition, the most able students could be barred from entering a career in music by parents who viewed it as beneath their status. We will never know how many potentially great musicians never managed to cross this hurdle. Marriage to a musician was also viewed as a downward social move. There is probably some justification in that: not only was there a servant type status but the income for the majority of musicians was not impressive. The income problem still remains for the majority of working musicians (pleasure in the music is only a partial compensation).

The reality for most musicians (i.e. not the big-name superstars) is that they are indeed in a servant type paid employment. It may be that they need to teach to raise money, even if teaching is not their ambition—with many of the pupils they encounter, it certainly will not be musically rewarding. Alternatively, they may be orchestral members who have to incessantly travel and play music set by the programme director and conductor, some of which they definitely may not enjoy. These are clearly drawbacks to a musical career, but in reality they are probably not hugely different from those in many other careers.

From my own experience as a scientist (a career I definitely enjoy) there are equally great irritations of teaching students who do not work or who lack ability, but pleasure from those with ability who also work. A less appreciated fact is that scientific work continuously needs considerable research funding for equipment, people, running costs and (in both industry and universities) significant overheads for the organization. Raising this money involves many research grant proposals (i.e. a type of begging letter) and, even as someone relatively successful with this activity, it has consumed at least one-third of my working time. So, as for a musical career, we accept compromises between the tedious and the rewarding aspects of our science.

In addition to the social class problem of music as a career, the image generated by the superstars such as Paganini, Chopin, or Liszt did not fit well with the general image of an suitable activity for a male. But female professional composers and performers were far fewer, and they had a different image problem. Not least, the prevailing thought assumed it was not possible for women to write great music. I have many CDs which emphatically say this is not true. For the men, musicians were often described negatively as Bohemian, undisciplined, sensitive, effete, or unworldly, and therefore music was not the lifestyle that one wanted one's sons to pursue. Clearly the thinking was and is inconsistent because in Britain many fine choirs and brass bands were formed by miners and factory workers, where the effete image of musicians was definitely not relevant. However, this was relaxation music, not career music. The same attitude towards a career that does not have an image of a macho or 'manly' identity persists for male ballet dancers. The reality is that even the chorus members are probably far fitter and more athletic than men in most other professions.

Perhaps the determination to become a professional musician, despite the social and cultural barriers and the image problems, acts as a filter to select the most dedicated and, hopefully, the most able.

The Relevance of a Changing Landscape for Earlier Music

To summarize, the dynamic landscape of music with developments in musical styles, types of instrument, concert halls, and the emergence of presentations driven by key soloists and ever larger orchestras has inevitably driven earlier music into a different, and often minor, perspective. There is a clear pattern, that influences from other nations and continents has totally altered our appreciation of music at all levels, and so now we have an enriched range of music and styles to enjoy. While this evolution of our musical knowledge means that we can only understand it from our present viewpoint, it does not inhibit us from trying to recreate earlier works, by perhaps playing them with period instruments and with intelligent guesses at the former musical styles. Such attempts at retrospective concerts cannot ever offer the actual conditions of performance, and certainly there is no possibility that we can match the audience understanding which originally existed. In many ways this is the same as for the groups who play at battle re-enactments. It may be fun, certainly less dangerous than the original, but it is clearly just a simulation.

For most readers these thoughts will be interpreted as perhaps comments on baroque or music from several centuries ago. However, our views are continuously changing, and since even music of the 20th century was scorned when it was first performed (e.g. Saint Saëns and Stravinsky) we should reassess what is historic. Listening to good-quality earlier recordings has been feasible for 50 years, and yet in many cases we hear totally dated performances (even by musicians who are still alive). On a broader view of such changing taste we only need to look at cinema films and musicals to realize that we cannot savour them in the same way as did the original audiences.

If you doubt this speed of change, then watch how areas such as radio or TV comedy programmes have altered well within living memory. Alternatively, try a general knowledge crossword from 20 years ago. Change in perception is inevitable, this is life—we are evolving.

CHAPTER 4

SIGNAL PROCESSING BY THE BRAIN

Sensing and Survival

The animal kingdom, as we know it, primarily relies on the two key senses of sight and sound, both for survival and communication. Creatures that lack either of these facilities do exist, but they mostly live in specialized environments where other sensing skills are used as alternatives. Heat-seeking snakes, sonar systems of bats, or pressure movement sensors of sea and underground predators all show that alternatives are viable ways of finding food. The sense of smell is equally important to them. This was once true for us, but for humans, particularly those living in urban environments, our senses of sound and smell have both become seriously degraded. We now have detection ability which is far inferior to that of even recent ancestors, or those humans who still live in low noise level environments. For musical appreciation this is extremely sad and has resulted in changes in the types of music that are composed and performed. The dependence on high level amplification and electronic music via poor-quality earphones on the smart phone are just two obvious examples.

Sight and hearing are so critical to us that we tend to discuss them together and ignore the other types of sensors. Indeed, they are complementary, but very different. It is worth comparing and contrasting these sensing strategies as it may help us understand more about their signal processing by the brain, and how and why we choose to hear some sounds as noise but other fragments as pleasurable music.

Alternatives of Discussing Wavelength or Frequency

In the sciences of sound and light there are alternative ways of defining the colour of light or the pitch of a note. Both involve energy travelling through the air to us (albeit with two different types of vibration) and for both 'speed equals the frequency times the wavelength'. In some contexts, it may be easier to think in terms of wavelength, and other times in terms of frequency, and the speed for sound (and music) in air is ~1125 feet per second (343 metres/second or 767 mph). This is all the science we are going to need if we want to switch between wavelength and frequency in our musical discussions.

Light travels very much faster, so if we see a lightning flash, and hear the thunder about 5 seconds later, then we know it was roughly one mile away.

Sensing Light

The light we use for vision always comes from outside of the human body, with the obvious original energy supplier being the sun. Light energy arrives in little packets of light waves which we call photons, and over a small range of their energies we can detect them in the retina at the back of the eye. For historic reasons we tend to discuss light in terms of the wave type properties of the photons, rather than in terms of their energy. Either description is fine. If we look at a rainbow we say it goes from red to blue wavelengths and appreciate the spectral colours. Sunlight, at source, covers a very wide range of wavelengths, but our atmosphere filters out a great deal of it, and transmits just a narrow spectral region which is strongest in the green (plus some much lower energy photons out in the far red which provide heat). Evolution has exploited this limited spectral range, and the results are eye type detectors in the majority of animals that optimize sensitivity to the sunlight reaching the earth. Humans have been quite modest in their choice of wavelengths, and we use a narrow band which extends from roughly the blue to the red. The peak sensitivity in the green precisely matches the maximum light energy that comes from a combination of the original sun's spectrum and the filtering effects of the atmosphere. However, in addition to the light we detect as humans, the atmospheric window also allows through both some shorter wavelength ultraviolet light and longer wavelength red. These extra wavelengths are used by a wide range of other animals.

Since the sun is the common energy source, most animals include this same spectral range. For us, we have decided that detecting the ultraviolet would be a bad strategy as the high energy of these photons will destroy our retina (as well as causing sunburn). For short-lived creatures, such as insects or even many birds, their life expectancy is so brief that they can afford to ignore the retinal damage, and benefit from the extra information encoded by the ultraviolet (UV) light.

Did you know that starlings look very different from one another in UV light, but far less so with visible light? Starlings obviously do, as they use the UV in identifying one another. A slightly disquieting thought is that for people who keep birds in a house, or in a glass aviary, then we should remember that glass does not transmit the ultraviolet. This loss of the UV information for the bird is the equivalent effect for us to live in a house where we have coloured windows that block out the blue and green end of the spectrum.

We also ignore much of the longer wavelength red (infrared) that is available to us, for at least two reasons. The first is that the physics of the detection needs more energy than is carried by red photons, and at our body temperature there would be a lot of background heat type noise blurring the image. By contrast, snakes in the desert are cold-blooded and hunt hot rabbits at night. In the cold they can exploit the longer wavelength red signals, as they do not care about a detailed image. A second reason for rejecting long wavelength red light is that our flesh is translucent, just beyond the red light we can view. Longer wavelength sensors can see beneath the skin (indeed, we are actively trying to use this to image cancer and other features). If our normal vision included infrared sensing, then people might look like some of the Leonardo da Vinci anatomical sketches of muscle structure.

If you want a feel for the scale of the wavelength of light, a red photon wavelength is around one hundredth the width of a human hair (and blue light is roughly half that of red). By contrast, a sound wave at vibrating at 440 times per second (called hertz, i.e. 440 Hz), has a wavelength of ~2.5 feet.

Just as with sound, we cannot only discuss the sensors but also have to fold into our analysis all the processing that takes place in the brain. In terms of sight we say normal colour vision spans the range from blue to red and we discuss, enjoy, paint and colour our clothes etc. From a host of shades, the standard eye system can probably distinguish at least 200 different ones. This is a brilliant piece of signal processing, as our retinal sensors are using just three broad bands of colour analysers, with a sensitivity which falls by more than one thousand times as we move from green to the extremes of blue and red. Our basic colour discrimination is quite crude, but the brain compensates so as to define colour by the relative intensity of just three colour detectors. This small number is adequate for us. By contrast, some shrimps have as many as 20 filter sensors. The simplicity of the three-band human colour discrimination has functioned well for our survival, but it is totally different from the way we approach and separate the different noises of sound into many component frequencies.

A Sound Strategy

Sound is a totally different playing field because the source of energy can be us, other people, creatures, or events that are happening around us. We therefore have control over the wavelengths (or frequencies) that we can produce and choose to detect. Here the wave properties of sound are important as the oscillating sound pressure in the air is influenced by objects of a similar size to the wavelength. Elephants and whales are enormous and have detectors (and voices) that run to very long wavelengths and they both use extremely long wavelengths (very low frequencies) for communication. For example, elephants can hear down below about 8 Hz. This is a frequency where there is a strong signal from thunder. Thunder implies rainfall, and elephants can use this sound to detect where to look for water. An 8 Hz sound has a wavelength in air of about 14 feet and because an elephant's head size and ears are a reasonably large fraction of this wavelength, they can sense from which direction it is coming. In part, this is because intensity differences are received between the two ears. The low frequency regime is also exploited by whales because such low frequencies propagate very well in water, and it gives them very long-range communication.

At the other sound extreme, bats navigate by sending out sound pulses to detect reflections that give them the distance to objects in their flight path, or to use a range of signal wavelengths to analyse the spectrum and decide if the reflector is food or not. To navigate and avoid objects as small as a telephone wire, or detect insects as food, means they need to use wavelengths as short as a few millimetres. Not surprisingly their sonar system runs up to frequencies as high as two hundred thousand cycles per second (200 kHz or 200 kilohertz). This frequency is equivalent to a sound wavelength of 1.7 mm.

These extreme animal examples show that we fit nicely into the same general pattern. From our head size and from the movements in our throats that define the wavelengths and frequencies of human voices, we typically are very happy with speech and sounds around 150 to 1500 Hz. If we had only needed to hear one another speak, then we would not have required a wider frequency range. However, the noises from the rest of the world are crucial for survival, and so our limits of hearing may run from say ~20 Hz to 20,000 Hz when we are young. Just as with the eye response we do not detect equally well across the entire range, and the extreme ends are also sensed less efficiently, by a factor of thousands or more.

The total frequency range we can hear is far greater than that needed for speech, but the extra information hidden in higher frequencies allows us to more easily identify individuals, separate the sound of different musical instruments, or hear bird song, or even the bottom end of the notes emitted by bats. This is typical of the entire animal kingdom. Figure 4.1 emphasizes a broad link between frequency range used in communication (i.e. animal speech) and body mass. The effect runs over a weight range of more than a million to one.

For many animals it is difficult to distinguish the difference between sounds intended for normal 'speech' and those used as specific warning signals. These may equally be considered as speech because the different Meerkats' warning cries define the type of predator. However, many animals, including humans and apes, use warning screams far above normal voice ranges. Data for horses do not seem to fit the pattern, as their 'voices' seem too shrill. One might guess that this is in part because we have used selective breeding to increase their size. Truly wild horses from Mongolia (called Przewalski) are small and weigh 300 kg or less, but our domesticated small ponies weigh in around 380 kg, and the really big shire horses can be as much as 1000 kg. Therefore Figure 4.1 only indicates the pattern where 'normal' communication can exist and excludes the higher notes of warning cries.

Two other points that can be appreciated from the graph. Firstly, I have used logarithmic axes because I need to go from extremely small to very large values. Musically this is sensible as we double the frequency when we go up an octave, so automatically this implies we have a logarithmic frequency hearing response. Secondly, we need to remember that the full hearing range of both humans and animals is considerably

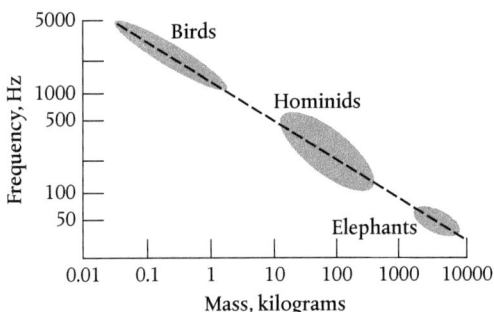

Figure 4.1 This shows that over an extremely wide range of body weight, different animals communicate at frequencies linked to their size. Data related to warning signals are not included.

greater than is used in 'speech' by any of these animal species. I find it intriguing that in this simple plot of typical communication frequencies and body weight are linked quite smoothly on this log-log plot.

A mathematical friend pointed out to me that this is a simple example of scaling theory. Frequency is proportional to 1/wavelength, (length^{-1}); mass is proportional to volume for animals of similar density, (length^3); so, on this plot, using logarithmic axes to cover a very wide range of numbers, we expect to see a slope of $-1/3$, which is exactly what I have shown with the dotted line. It means that big animals have low voices, and small ones are high pitched. The pattern flows almost smoothly from birds to elephants. Even with humans, we are not surprised that small children speak with higher voices than adults, and bigger adults generally have deeper voices than lightweight ones.

I also find it comforting that our best sound detection is in our speech range, as this is saying that speech must be more important to us than any of the other sounds that we might hear. Musical instruments cover a much greater frequency range than our voices, but we never expect to listen to music which is much outside of our voice range. It is of course true that double basses and organs play notes that are far below the lowest male voice but, as I will soon explain, these are never pure single notes, but instead a package of many frequencies of which we name the note by the lowest component, but often interpret it from the higher frequency content—indeed, we may not even hear the lowest note!

If we discuss the frequency of a note, such as that of a violin A string, then the lowest note it produces is at 440 Hz. However, even for such a simple string, there are weaker components at other frequencies. For this violin string the numbers are multiples of 440 Hz, so are at 880, 1320 etc. These extra notes are called harmonics (or overtones). The pattern of their spacing and their relative intensities define all the key instrumental sounds that we hear from each instrument. A term used to describe the package of harmonic notes is a formant.

For notes sounded above the speech range there are harmonics that are similarly higher than the named note, but we do some spectacular signal processing with the entire package, and then find features that relate to our speech range. Music written solely above our speech range might make an interesting experiment for a composer, but I suspect that it would have limited appeal, except as a gimmick.

The Role of the Brain

For both sight and sound we have pairs of detectors, and because they are displaced, the left- and right-side signals are not identical, but depend on direction and distance of the source. The two systems also differ in that vision only offers a forward view and we can consciously shut it off if we close our eyelids, or when we sleep. This contrasts with sound detection which is both continuous and gives signals from all directions. Directional features, and signal content only usefully emerge after the signals have been fed to the brain and processed for interpretation. In both sight and sound, we need to understand the complete sequence of collection, processing, and analysis, which takes place within around 20 milliseconds. It is an understatement that this is a major design challenge, but it explains why the brain is continuously functioning and uses roughly a third of our energy output. The two processing detector systems (light and sound) equally consume almost a third of our brain capacity. The signals need to be accurate, rapidly processed, and easily interpreted. The amount of input information is excessive and so the only option is to make a number of compromises. Any engineers who were faced with making such a design for a compact lightweight processing unit would rapidly realize that it is essential to make as many approximations as possible, that will increase processing speed, and minimize the fraction of the brain used for detectors.

With both visual and aural signals we operate from a minimum background level to intensities that are more than a million times greater. Compromises are to use a logarithmic intensity scale (i.e. we switch from counting as 1 through 10 to 100, 1000…etc. to units of 1, 2, 3, 4…). We also limit our working detection ranges. The compromised design approach is to say we want excellent responses only in certain regions (green light or speech), but are much less interested in signals further away from the core region. Therefore sensitivity to blue and red light, or to very low and high audio frequencies, reduces down by a thousand to a million times, relative to the efficiency for detecting the central green signals or the speech frequencies. We have totally focused our priorities on a very limited central region in each case.

Secondly, we basically do not care about absolute measurement of intensity, only how bright or how loud it is compared with other sounds. This is a very reasonable way of reducing the complexity of the detectors, as we never want to hear, at the very same time, an intensity sound range that spans the noise of a breeze up to that of a trumpet fanfare. Basically, we do not need the high precision of a physics measurement on a nice linear scale from one to a million. Instead we are happy with a relative scale that operates over a small intensity range. From this perspective we just have to recognize that one sound is louder than another sound, and perhaps be able to say it is twice as loud or four times as loud. Overall, this is a fantastic saving in computational power, storage, and energy to operate the brain.

Brain processing using the arithmetic of logarithmic scales is what we intuitively use when there are a wide range of numbers. A range from 1 to 10 million just looks like 1 to 7 and going from 2 to 3 etc. just implies an intensity doubling. Biologically brilliant as we have scaled down our computer system by ~100,000 times. A head size brain is then possible. I mentioned that the processing of sight and sound consumes about 30 per cent of the brain capacity, with a power demand of about 10 watts for this task (even with the logarithmic scale route). Any linear attempt would have been impossible in both size and power consumption.

This selected range approach is immediately obvious and familiar because in bright sunlight we can only see brightly illuminated things but at night, when the average level is much lower, we can see the night sky and stars even though they are a million times less bright than full sunlight. The same is a familiar feature of sound detection. On a really quiet day we can hear a breeze moving the leaves in a tree, but in a busy city street we may have difficulty hearing someone speak. Musically this has immense implications as we cannot simultaneously process very soft and very loud sounds. The soft ones will be lost. In an orchestra the various instruments differ greatly in power. For example, a trombone will totally swamp the sound of a single violin. Excellent news for violinists, as the orchestra needs to employ a dozen each of first and second violins in order to double their perceived power level.

This is a factor that has not been fully understood by many composers, who still write notes in subtle weak sounds that are drowned out by a brass fanfare. Overall it is the limitations of packing a powerful processing unit (the brain) into a small volume with low energy demands that has defined, and limited, our visual and sound responses.

We use the same mathematical trick in describing the natural phenomena that span very large dynamic ranges. For tsunamis, hurricanes, tornadoes, and the intensity of earthquakes, we use logarithmic scales where every power increase by ten times notches up one unit on the scale. The difference between an earthquake of scale 3 and 5 is a factor of one hundred in terms of the energy released.

Musically we already use logarithmic scales for the tuning of the notes, because when we go up an octave, we have doubled the frequency. While writing this I realized that a piano keyboard is an instrument that visually displays a logarithmic scale of frequencies, so this is a tangible example of such a scale, and one which we accept without any hesitation or fear of the use of logarithmic mathematics. In precisely the same way all those who play string instruments (cello, guitar etc.) double the frequency (i.e. go up one octave) by halving the string length. So musically, logarithmic scales are familiar territory.

The Design Challenge for the Ear

High-grade human hearing covers a frequency range from ~20 to 20,000 Hz. Here we differ from the treatment of colour using just three colour detectors. Instead we want to be frequency precise to better than about a quarter of a tone in the range of speech. The real situation is yet more challenging because signals do not come in isolation, but are a mixture of many frequencies, from each and every sound source. Further, there will be many sources providing signals, and we must discriminate between them. Voices and musical instruments never generate a single frequency note. Tuning forks can be a good approximation to a single frequency source, and cover a very limited frequency range, whereas a voice will have continuous signals over most of our hearing range, including notes that are below the ones that are intended! Our key challenge is to identify individual voices and instruments (or dangers) within a few tens of milliseconds. This is despite the fact that the sound pressure on the ear fluctuates at an extremely low value. The eardrums are actually displaced by sound pressures far less than would be generated by the weight of a gnat dancing on them.

The benefit of a long evolutionary history is that the problems have broadly been solved. Our bodies have made the sensible expected compromises that detection is optimal over just a limited working range but the components can be damaged by exposure to high intensity, in addition to fading performance with age. Less obvious are a number of quite unexpected oddities in our responses that most of us would not have guessed could exist. Some of these have important consequences for listening, and also for appreciating music, so they are worth a mention. I will attempt to explain how they come about. 'Attempt' is the key word, as there is not total agreement on the cause of some of these peculiarities. In part it is because no two people have identical hearing (or identical ear responses) and the brain processing develops according to our experience and training. Once again this implies that music and sound analysis are highly individual. With clear differences between test subjects it is difficult to produce results that allow a unique description of subtle effects.

Most textbooks and articles would, at this point, show a diagram of the ear with all the bits labelled. Without very careful perusal, these diagrams or photographs are not very helpful. Instead, I will offer sketches in Figure 4.2 that are easier to relate to our needs, and also should offer more insight into some of the oddities and demands. I have ignored the semicircular canals, which are important for balance, and shown the cochlea as though it were straight.

A real ear has an external section (the pinna) which directs sound into a narrow canal leading inside the skull. The external part gives preference to sounds from in front, or from the side, and focuses in more sound than would be picked up by small aperture canal alone. Big ears are therefore likely to offer better sensitivity (e.g. bat ears are proportionately

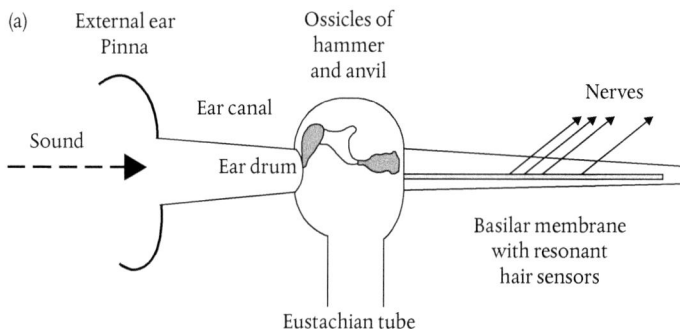

Figure 4.2a An engineer's view of the ear, with sound wave energy conducted in via the canal from the external ear. There is then a section to efficiently couple motion in air to wave motion in a fluid in the inner ear; and a membrane region that is sensitive to different frequencies. The principles are simple but the actual responses are remarkably complex. Finally, the output signals from the membrane go to the brain for processing. What we hear and interpret is dependent on all these stages.

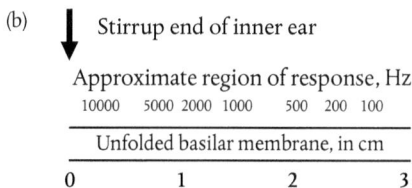

Figure 4.2b In this expanded view of the region of the basilar membrane it appears that the frequency regions are mapped almost logarithmically along it, with high frequencies near the ear drum end and low frequencies at the far end. This is the same pattern as for a piano keyboard but with high frequencies near the ear drum end.

very much larger than human ears). Early ear trumpets for those going deaf did just the same and collected more sound energy. Ear shapes are quite diverse and have additional functions of appearing attractive (or not), and are convenient for supporting spectacles, or attaching ear rings. The actual business of detecting the sound commences at the end of the canal, where the fluctuating pressure wave (the sound) pushes on the ear drum, and minutely makes it vibrate in and out into the canal.

We need to couple this sound energy into the fluid inside the inner ear (the cochlea), but if we tried to do this by the movement of the eardrum the huge (thousand-fold) difference in density between air and fluid, means virtually no energy would transfer. Instead the sound energy would just be reflected. The role of the middle ear section is to overcome this impedance mismatch. It includes a brilliant little moving hammer, anvil, and stirrup structure made with the smallest bones in the body. This sets up pressure ripples of the same wave character into the liquid-filled tube of the inner ear. The anvil structure does not merely overcome the energy loss from reflection at the eardrum, but actually amplifies the signal onto the smaller inner ear section by factors in excess of twenty times.

The anvil system has an important second function and that is to stop us being permanently deafened by hearing loud noises. There are a set of control muscles that can stiffen and reduce the coupling efficiency, and so protect the inner ear from too strong a signal. For steady signals this works well, but it takes about 20 milliseconds for us to detect a signal and send a return message to tighten the anvil muscles. Unfortunately, if there are sudden explosions, the response is too slow. The muscles cannot reset fast enough, and we can end up with permanent ear damage. If we know or see that there is to be an intense sound, we can anticipate the muscle control. This even happens in a car crash where our anticipation reduces the chance of being deafened by an exploding air bag. Precisely the same problem exists in bat sonar (and the radar equivalent). The bat needs to fire an extremely high power directional pulse of sound, and then time how long it takes for the very weak reflection to return, in order to assess the distance. The output pulse would destroy the ear, so during pulsing signals out, the bones are disconnected from the eardrum.

In the next phase within the inner ear the signal travels through the fluid over two continuous sets of hairs that have different stiffness, and natural oscillation frequency (effectively the inverse of a piano, but with many more keys). When a hair resonates it then sends an electrical signal to the brain. For many reasons, the hairs are mounted on a support which is slightly flexible but which becomes stiffer for higher frequency. (Think of the detector mounted on a hard rubber sheet, or the bristles on a rubber-based hairbrush.) The high frequency end is near the ear drum.

At low intensity, the hairs neatly and selectively resonate, to record the incoming signal. At low power, they are sharply tuned and are independent of their neighbours. However, at high intensity the supporting base membrane starts to flap, which in turn causes adjacent hairs to be stimulated. The brain is therefore fed signals over a frequency range, rather than at the exact frequency of the sound. Because the support sheet is stiffer for higher frequencies the neighbours on the lower frequency side respond more than those on the high side. I have sketched the effect in Figure 4.3. The frequency distortion is greater towards lower frequencies, so we hear a less precise note that has been flattened in pitch.

It is easy to demonstrate the effect by hitting a piano key hard to produce a loud sound, and then hold the key so the piano note can continue to resonate as it slowly fades away. As the intensity decreases we appear to hear the frequency rising slightly. This is the prediction from the effect that I am describing within the ear, caused by a stiffness that increases at high frequency. The piano demonstration may be cheating a little, as a similar intensity response might occur within the instrument. Nevertheless, the effect we perceive is that the sound seems to drift out of tune with high intensity. Potentially this is a real problem for singers, who hear themselves at high intensity, whereas a more distant audience will never hear the same power. So, the mental tuning correction for the singer will aim for a higher note than is needed for the audience. Indeed, many critics say a number of powerful sopranos drift out of tune on high notes. This is probably because they internally hear a very intense sound, whereas by the time it reaches the audience, the power levels are

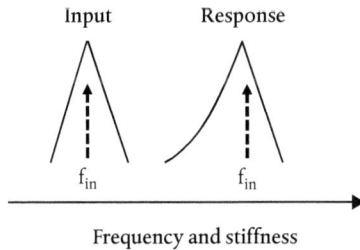

Frequency and stiffness

Figure 4.3 The symmetric input describes a frequency spread about a central value of f_{in}, but the membrane varies in stiffness so, at higher power levels, it vibrates asymmetrically about this central frequency. The effect is that the base support movement stimulates the response hairs more on the lower frequency side. Overall this produces shifts in both the peak value and the average frequency value (i.e. we perceive a slightly lower pitch note).

more modest. Knowing how to match the internal and exterior frequencies is difficult in this situation.

In the past I had noticed the effect from the piano, but initially assumed it was my imagination. One may also consider whether the effect is apparent when using pre-recorded accompaniments or for Karaoke singing. Does the volume level influence our own tuning?

The same response characteristics of the ear are equally obvious if we hear a pair of notes that are closely spaced. Here the patterns are shown in Figure 4.4, for three alternatives. In the first, the pair are at equal low intensities. This is fine because we hear both of them correctly in tune. However, if the lower one is more intense, then we hear both notes with the lower one slightly flattened, and the weak high note still in tune. Conversely if the high note is intense it will flatten and swamp the weak low note. The apparent maximum sound to the brain will be interpreted as though it is just one intermediate note.

Again, this may seem surprising, but in fact we frequently observe this response. In a crowd with a lot of background noise, we have problems distinguishing people with lower voices, whereas someone with a shrill

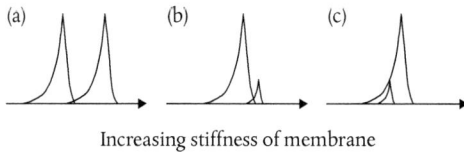

Increasing stiffness of membrane

Figure 4.4 If we generate a pair of signals that are equal in intensity and symmetric about their central value, this is not precisely the response transmitted from the ear to the brain. The changing stiffness of the membrane skews the signal to lower frequencies. Three examples are for (a) equal intensities for an input pair of weak signals. The more flexible end of the membrane responds to high loudness. So even in this case there is a broadening of the frequencies transmitted to the brain, and the broadening is on their low frequency side, as the lower frequency hair detectors are stimulated on the more flexible section. The signal is therefore no longer symmetric in frequency, nor is it precisely mimicking the input frequency. (b) One loud signal with a weaker sound on the stiff side. This allows the weak upper sound to be heard, and we hear a pair of notes. By contrast, (c) is for the same frequency pair, but with the more intense sound at the higher frequency. The flexibility of the base support spreads the vibrations to the flexible side (where there is the weaker signal). The net effect for (c) is that the weak sound is masked within the envelope of frequencies transmitted to the brain.

voice will carry above all the others. This is called masking. The distortions generated within the ear can be exploited by a violin soloist, and help with a problem when the soloist is playing the same passage as a large orchestra. Rather than just fade into the background of the orchestral players, a highly skilled performer can be just slightly sharp (slightly!). Then the presence of the soloist will be evident as distinct from the orchestral players. This is a dangerous strategy as an excess will appear out of tune. Precisely the same techniques can be used by singers, but this is even more difficult as the sung notes are less precisely defined compared with those from a violin string.

Figure 4.4 needs some cautious reading, as it is not intuitively obvious. It says that the notes we hear can drift in frequency if the intensity level changes. Ignoring the details, it says (i) the note we believe we hear can change with the volume, and (ii) if there are two notes of different intensity the lower one can easily be masked. The effects are real, but surprising.

As already mentioned, the same situation is equally obvious in a big party with a high sound level. People with slightly lower voices than average are masked, but a shrill voice will emerge above the background. Masking effects can occur in musical situations, and for example we find it much easier to follow the tune being sung by the highest voice (or instrument) compared with the parts from the lower ones. The overall sound certainly benefits from the lower parts but separating them from an ensemble is far more challenging than listening to the top notes. This is not just a classical music situation but is equally obvious in pop and other music; it is one reason why successful male pop stars often sing with a high voice, despite the fact that there is a female preference for men with deeper speaking voices.

Just to add confusion to this suggestion that the most important tune will be in the high voice line, one can look at early church music and motets. There the main line was often given to the tenor range. In a church choir this might have been below the treble sounds of choirboys. Indeed, the name tenor is derived from the fact it holds the tune (from the French *tenir*, to hold).

The final related effect is that a singer, as distinct from say a cellist, is producing the notes within the body and all the bone and tissue transmit signals directly to the ear structure. I did not pick the violinist in this comparison as there is some direct energy coupling via the instrument via the chin rest to the bone structure. I have experimented with this and it appears that for the violinist there can be some minor tonal differences between a loosely held violin and one that is firmly gripped. Consequently, for the singer, the power distribution along the line of detector hairs is quite different from that heard by the ears of an audience. Singers always say that they never sound on a recording the way they thought they did during a performance. This is not imagination or wishful thinking, it is true. Unfortunately, the oddities such as the masking effect I have been describing also mean that the frequency of the note they 'hear' according to their brain (especially at high power) may not be the same as the note they are actually singing. The singer's brain will control the pitch and may actively mislead the voice because of complexity in the vibration patterns within the inner ear. The problem should be less for pop music performance where the singing volume can be low, as there is electronic amplification, and many singers have an audience feedback headphone. Big stage opera singers are potentially more likely to be off with high power top notes, even though they believed they were in tune.

The Frequency Dependence of the Ear

I have now mentioned the key facts that our detection system runs over a wide frequency range, it is optimized for speech type frequencies, is poor at much lower or higher frequencies, and we sense the intensity and frequencies with logarithmic responses. All this information can be captured on one diagram, but instead I will give two versions of it in order to relate it to speech and music. Figure 4.5a, contrasts our response to detecting different frequencies, from threshold to pain level. The data are for those with good hearing. The frequency scale along the bottom axis is effectively that of a piano keyboard plus an extra octave (i.e. the keyboard is a logarithmic display of frequencies of the notes).

The contour lines on this map show how we assess different sound intensities from very quiet (*ppp*) to very loud (*fff*). The details are not too important, and differ considerably from person to person. I have labelled the intensity axis in units called decibels (dB), other texts may use different sound units, or quote the vertical axis in pressure units. Decibel units are useful as the only key feature for us is to note that a change of 10 dB is perceived as a doubling of the sound intensity.

The responses shown in Figure 4.5 vary between individuals, but for all of us the pain threshold is similar. The rest of the diagram will differ with age and/or exposure to loud sounds which can easily cause a permanent hearing loss. A hearing loss of 30 dB, which is typical for older people (i.e. typical of many of the classical concert hall audiences) is equivalent to a factor of 1000 in terms of intensity loss. The bonus of a brain working with a logarithmic response is that we only consider this as though it is 8 times quieter.

For those interested in actual power levels our sensitivity can span a range of over 100 million in terms of power level, the range goes from about 10^{-13} up to 0.1 watt per square metre.

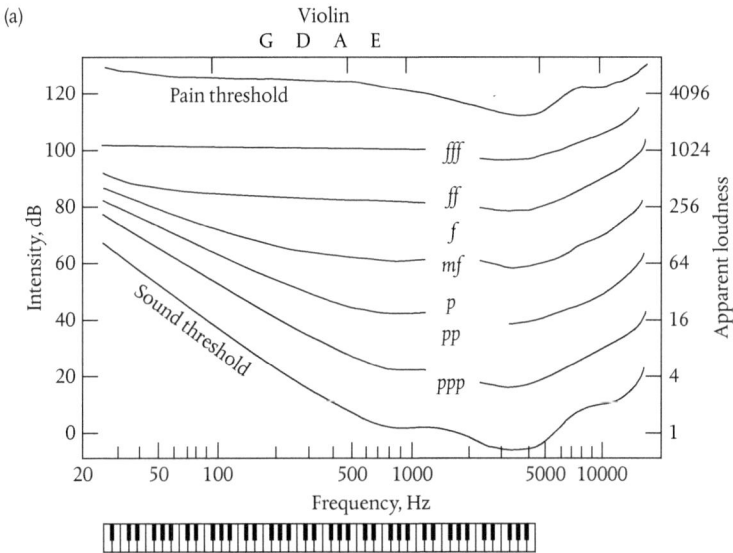

Figure 4.5a An overview based on patterns of good hearing response when young. Attempts to make such measurements are difficult and variable, and in detail, examples differ with vintage and technique. When young we have a wide range of frequency coverage and dynamic intensity range. The power units are in a logarithmic scale of decibels. The right-hand axis indicates how we interpret this in terms of apparent loudness. The frequencies of the violin strings G, D, A, and E are indicated for reference. The logarithmic frequency axis is the same as that covered by a piano keyboard but our hearing extends to higher frequencies by just over an octave.

(b)

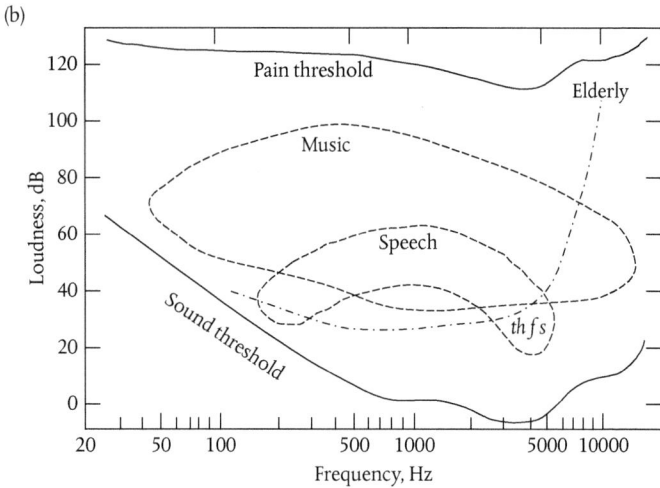

Figure 4.5b An indication of the frequency ranges covered by music and speech. The banana shape for speech shows sounds such as *th*, *f*, and *s* are at high frequencies and so are lost through even modest hearing loss. The coverage of speech and moderately loud music are noted. The fundamentals for speech are typically from about 120 Hz (men with bass voices) to 5000 Hz (women and children). Our voice formants contain higher frequencies, so we need response to at least 5000 Hz. However, the threshold sensitivity response of an elderly person is much poorer at higher frequencies, so speech can seem blurred. Even worse responses often occur for those (of any age) who are exposed to loud noises or high-volume music.

The second version, Figure 4.5b, offers some idea of values for the lowest components of sounds used in speech. The typical bottom ranges are in the region of 100 Hz (men with bass voices) to 500 to 1000 Hz (women and children). Hearing only the fundamentals would not help us distinguish between people, so we need a speech range that includes many of the higher harmonics of our speech. The details of the harmonic content are specific to each person. The upper limit needed for hearing clear speech is usually near 3000 Hz, but higher for music. As a reference, the frequencies of the violin G, D, A, and E strings are indicated on Figure 4.5a. Responses are very variable between people, and change with exposure to noise and with age. An indication of the threshold loss with age for an undamaged ear (i.e. not from noise) is sketched. Negatively, it shows why we typically lose sensitivity to bats and bird calls as we age. Positively, shrill sounding instruments may appear to sound sweeter—indeed, I mistakenly thought my violin playing had improved as I grew older!

Hearing Quality from Historical Records, Damage, and Aging

Hearing is one of our most highly developed senses and our ability to recognize and distinguish sounds has meant that not only do we use it for survival and communication, but with 'civilization' and urbanization, plus progress in technology, it may mean we do not appreciate sounds as once was routine. Almost certainly the average person in the modern 'advanced' world will have inferior acoustic ability and ear response compared with those who are living on undeveloped Pacific islands, or in South American jungles. In order to overcome the masking effects from background noises, the same species of birds sing louder in our cities than they do in the countryside. Our loss is difficult to imagine because we grow accustomed to living in a background of high noise levels. There is also the very sad fact that exposure to loud noises and high backgrounds means that within a relatively short period of exposure we not only have some temporary deafness, but our overall sensitivity is significantly reduced. Damaging noises for people working in factories is not unexpected, but most studies of the subject show that working in high-noise cafes, travelling on the New York Underground etc. can all produce major permanent changes over relatively short periods of exposure. The quoted numbers are often surprisingly large, for example effectively halving our dynamic intensity range. A halving does not seem serious but it is a logarithmic scale so halving the number of *decades* means, in power terms, that we have raised our minimum by almost 1000 times! Music played with electronic amplification, whether in pop concerts or via headphones, with too high a volume setting is equally damaging. Many pop stars and disc jockeys are partially or seriously deaf as a result of their exposure to high intensity music. Unfortunately, even a short exposure will temporarily cause some deafness, and the instinctive reaction is to then raise the volume, and hence cause permanent damage.

This scale of hearing loss with time is difficult to quantify as no serious measurements were being made until around 1870 by Töpler and Boltzmann, who looked at frequency response, and later by Rayleigh in 1877 with studies of threshold sensitivity data. Earlier clues come from older works of literature that include the fact that people could shout and communicate over fields and farmland on distance scales of a mile or so. The numbers are so large we now assume the distances must have been written by mistake, and never realize that the change is in our hearing and general background noise. Nevertheless, older countryside friends support the claims. I also remember sitting reading a book and hearing an unusual noise. When I looked up, it was from a large spider striding across my wooden floor, so at that stage I also had a good lower sensitivity threshold.

As a physics student I read an old standard textbook from 1940 called 'Acoustics' by Alexander Wood. In it he quoted examples of experiments from the 19th century to find the frequency range that was detectable. The examples he cited not only showed that healthy subjects were able to hear frequencies as low as ~20 Hz and as high as about 20,000 Hz, as we might expect for a young person today, but the high frequency limits are intriguing. Wood gives 19th century data by König, who claimed that at 41 he could hear up to 23,000 Hz and this fell to 20,480 by 56 and was still a surprising 18,400 Hz at 66. These numbers are spectacular because for a modern population it would be exceptional to find *any* adult city dweller who could ever hear such high notes or have such an excellent response at 66. We may now think 66 is still 'middle age' but in the 19th century it was typical UK male life expectancy, and indeed 65 was set as the age limit for pensions because the government assumed it would not be expensive to fund.

If we look at aging effects on vision then there are some interesting examples of how they influence the creativity of painters. There is the classic example of Monet, who slowly developed cataracts but he frequently continued to paint a view of a small Japanese style bridge over a pond in his garden. As the cataracts developed his painting style became more blurred, and progressively he lost the short wavelength blue tones. Eventually he produced only blurred red pictures. He was then fortunate to have the cataracts removed, good vision and colour reception were restored, and he was horrified by what he had painted.

I wonder if we could identify similar changes taking place in sound for compositional styles as composers grow older and lose high frequency response. Do they write with fewer high notes? Would they revert to earlier compositional patterns with the addition of good hearing aids? Making a clear identification of such an effect will be challenging as for a really great composer, Beethoven, his internal imagination as to how music would sound was so spectacular that he produced masterpieces long after he was deaf.

Hearing Loss, Hearing Aids, and Their Relevance to Music

In previous generations, hearing loss for young people was invariably related to extreme noise conditions of factories or warfare, and it was also assumed to be a natural result of aging. The noise level in factories may have been reduced but for the current generations, hearing loss is all too easily self-induced by exposure to high level noise in discos, rock concerts, or through headphones with the sound level set too high. This can be very obvious even from the sound levels one hears from cars driven by young people. Therefore, while older reference texts cite hearing performance of extreme sensitivity in terms of low volume, and ability to hear up to 20,000 Hz, these numbers are totally unrealistic for the 80 per cent of the population who now live or work in cities, or have attended discos etc. (Note city birds have to sing twice as loudly as their country cousins.) The modern reality is that, for a young adult living in a city their upper frequency is rarely above say 15,000 to 16,000 Hz.

Despite such widespread hearing damage, the image is that hearing aids are primarily for older people, and this is from normal deterioration; rather the actuality is that it may have been linked to life styles with excessively loud sound levels. Numbers are hard to quantify, but a BBC report in 2018 for mature people claimed 40 per cent of over 50s had some loss, and 70 per cent of those over 70 were noticeably afflicted.

Avoiding exposure to excessive sound levels is not easy. From bitter experience I have been to wedding receptions etc. with discos, where the sound levels meant that speech was impossible, and with anyone using a hearing aid the volume will be lifted far above an intelligible threshold. As mentioned, the high percentage of disc jockeys and rock stars who have permanent hearing loss makes this point very clearly.

Regulations have varied, but in recent years one can cite discos typically operate at 100 to 110 dB, and being close to a loud rock band can be nearer 110–115 dB. Anyone with, or without, a hearing aid will thus not merely be exposed to sound above the pain threshold, but it will induce rapid hearing loss. This includes the musicians of all genres, including classical orchestral performers, seated in front of the brass, who often have impaired hearing. Legislation to reduce this is now in place and it will be effective because orchestral players have successfully sued as a result of the damage to their hearing.

Quite typically, as sketched on Figure 4.5b, in the mid-speech frequency range the losses from aging may be say 20 or 30 dB. Remember that 30 dB means an immense change of one thousand times in real power, but because our brain processing runs on a logarithmic response we perceive a doubling (or halving) of sound intensity for a 10 dB change. So 30 dB is eight times louder or softer. If this were uniform across our hearing range it would be simple to just amplify by 30 dB to offset the loss. However, as sketched on the figure, the loss invariably increases steeply at higher frequencies and we may be effectively deaf by 8000 Hz.

There is thus a small technological problem that a uniform power boost across the entire frequency range will not greatly help the high frequency end of music, but instead we need an amplifier response that changes with frequency to compensate for the performance of each ear. A second requirement of the electronics is that if we are listening to high power levels of music (e.g. in a disco or a brass fanfare) then the power level may already be near the upper limit of our intensity range. Further amplification is then disastrous as it will raise the sound level to where it is both painful and can cause deafness. This upper region on the diagram does not change because we fail to hear low power. Especially in classical orchestral music the intensity levels can range from a whisper to intense. A well-designed hearing aid must therefore compress the amplification in order to stay below an upper limit. This is not too difficult, and indeed compression of the intensity range is used on most CDs and radio stations. Neither offer the dynamic range which one would experience in a concert hall.

Hearing aids require complex engineering and are expensive, but it is not always appreciated that the requirements for music may differ considerably from those used for normal conversation. Indeed, most hearing aids only boost power up to 6000 Hz (i.e. the top end on normal speech). This is valuable as it adds the sibilant sounds of *sss* or *th* etc., which means more clarity of speech. For the general public to be told that your hearing range has reduced from 16,000 Hz down to 8000 Hz is a traumatic event as they will think they have lost half of their audio range. The numbers are correct but the reality is very different. From 8000 to 16,000 Hz is actually just one musical octave. To put this loss in perspective we need to consider how many octaves we can still hear. Thinking in terms of notes on a piano is helpful as it typically has about seven octaves from A_0 to just above A_7 (i.e. 27.5 to 3520 Hz in equal temperament tuning). Even with the response of the impaired hearing plot on Figure 4.5b we will detect another octave up to A_8 (7040 Hz). In total we are sensing (albeit less than perfectly) a range of eight octaves. The *good* news is that losing response from ~8000 to 16,000 Hz is indeed a large frequency range, but it is just one octave of the nine octaves we could have heard when young.

The musical problems of hearing aids become quite complex as a minor loss in the lower speech range means we hear some sounds directly, with or without the amplified signal. Amplification to compensate for losses at high frequency require detection, electronic processing, and delivery to the ear. This takes time and often results in a delay of a few milliseconds. Hearing both direct and boosted signals that are not totally synchronous can cause a reduction in sound quality, and in some case a slight sense of unease (particularly if we can see the sound source). Typically, a delay of just a few milliseconds is acceptable, but if it rises above 10 milliseconds it then may be described as sounding metallic. Longer delays are of course a disaster as they will become an echo. It may seem strange that short electronic delays of a few milliseconds are important, but if there is a note at say 500 Hz (mid-speech) then the pressure wave on the eardrum is oscillating at this frequency every 2 milliseconds. The hearing aid always delivers a boosted sound which is delayed, both because of the electronics and a sound tube of some centimetres length. The consequence of this delay is that it will be out of phase with the original pressure wave. This means there is some frequency-dependent cancellation, and we hear a change in sound quality. Our brain processing of audible signals is brilliant, so we notice it, but can adapt.

Perhaps the final question to ask is why one does not hear a greater discussion of the musical difficulties involved with compensation for hearing loss. My view is that for use in conversation the range of power levels and the frequency range are both relatively limited, and the bonus that is offered by a hearing aid outweighs the distortions that exist. In general conversation we are keen to hear the content, but not concerned with the detailed sound quality. For music one must be honest and admit that listening, for the majority, is in an environment where there are other noises and/or we may not be focusing in detail on the music but be at the computer, reading, travelling on a train and listening via a smart phone, or have further background distractions. This is a widespread behaviour but probably one that undermines our musical appreciation. By contrast, in a concert hall where we are likely to be fully concentrating on the music, the inherent imperfections of the systems are apparent to many. 'Fully concentrating' is somewhat idealistic. For the Chopin 'One minute' waltz, or a typical song lasting five minutes, then this is quite possible, but for something as extreme as the 8th symphony of Bruckner, which lasts around 100 minutes, even the most dedicated musicians will be severely challenged.

The use of standard hearing aids by musicians apparently is quite typical in that they use them for conversations, and they like the brighter sound generated by the boost of high frequencies. At concerts in large auditoria, where they are not close to the stage, then here again they are beneficial. However, in small music rooms, or when actually playing, they may not use them, and have a standard criticism that the mid-range sound is blurred and fuzzy. If you need a hearing aid for music then be sure that the audiologist understands this, as you may benefit from a different software programme than a standard issue device. A final comment is that when listening to a scratchy violin tone, one may hear a sweeter version with loss of high frequency response!

CHAPTER 5

INTERPRETING COMPLEX SOUNDS

The Brain and Signal Interpretation

I have mentioned that the inner ear sends electrical signals to the brain, and used a simple case of isolated notes. In practice clean, single-frequency notes are extremely rare, and among the few examples are the notes produced by tuning forks. A tuning fork marked as an A at 440 Hz will indeed produce that frequency, but virtually nothing else. Our need is to have music and speech with identifiable sounds. With tuning fork purity, everyone would sound the same. Instead, our voices and all instruments produce many notes (with varying decay patterns) which are in addition to the frequency with which we have named the note. For some simple instruments such as the strings (violins etc.), woodwind (flutes, clarinets etc.) we understand the acoustic properties of the instruments and can both predict and partially control the mixture of higher harmonics. Such control separates the skilled makers from those who are more mundane. The harmonic content is so important that a mass-produced violin may sell for as little as £100, but for an instrument from a top maker, such as Stradivari, the price will be measured in millions. We care about these subtle distinctions, as our brains are capable of ana-lysing the differences in the time that it takes to *start* playing a note. We take some 20 to 40 milliseconds to produce a detailed identification of the fundamental frequency, the harmonic content, and the type of instrument (or person) who is producing the sound. Impressive!

We then only need a few notes (or words) to identify a speaker and scarcely more to recognize familiar tunes or identify a composer and vintage of the composition! Not only can we do this in isolation, but we can do it in the presence of background sounds, a symphony orchestra, or many speakers. An old but familiar CD such as that of the Three

Tenors (Carreras, Domingo, and Pavarotti) has three excellent tenors singing, both separately and together, in the same frequency range (i.e. the same notes). Despite this, we have no difficulty identifying their individual parts. This is because each of them has differences in the design of their vocal chords, sound production, and body shape, etc., so the harmonic pattern of notes (the formant), frequency spread, and speed of attack and fade all differ.

Speech examples are very complex, so instead I will take an easy option and try to explain the formant pattern for a note from a violin string. In a later chapter on the voice I will show some formant data for singing.

The Harmonic Content of Notes from an Isolated Violin String

From the million-pound Stradivarius violin we might expect some high technology, but the reality of violin design is very simple. There is a stretched string with fixed ends (at the bridge and the end of the fingerboard) with an energy source of a moving bow. A good bow will use white hairs from a horse's tail (the white horses of the Camargue are favoured). The surface of the hair is naturally serrated but to give it even more friction we make it sticky by rubbing on rosin, which comes from the sap of a tree. Because it is sticky, when we pull the bow across the string, the string is steadily moved sideways. However, the frictional force is limited, and when we hit the limit, the string flies back very fast, so this produces a sideways movement pattern that looks like a sawtooth. The frequency of the sawtooth is fixed by the resonances of the string. If we shorten the string we get higher notes. Also, as expected, the fundamental note that appears is the note that we intended to play.

Rather than describe the motion in terms of string movement we want to think in terms of musical notes. For the pure tuning fork note the movement pattern looked like a nice sine wave (i.e. smoothly varying from side to side). To simulate a sawtooth movement at frequency f, we need to add together a group of sine waves with frequencies f, 2f, 3f, 4f etc., and we need them to have different amplitudes that are in the ratios of 1:1/2:1/3:1/4 etc. This is not quite the end of the story because our ears do not respond to the amplitude of a wave, but to the pressure, which

depends on the square of the amplitude. Therefore, the harmonic intensities heard from a bowed string come out with a ratio pattern of
$1:1/4:1/9:1/16$ etc. We can either consider the mechanical displacement of the string or, more conveniently, think in terms of a set of harmonics of different intensities. The two views are totally equivalent, and
mathematicians will term them a Fourier transform of the same event.

The sketch in Figure 5.1 says the same thing via a diagram. Note that
although the range of intensities from the fundamental (i.e. the first harmonic) to the 10th, is one hundred times, we actually would only *hear* the
higher note as seeming to be just a quarter as loud, *if* we had the same
sensitivity to both notes. Normally we do not (Figure 4.5a) and our perceived change in intensity depends on where we are on Figure 4.5a,
which shows how the importance of higher harmonics depends on
both the starting fundamental and the frequency response curve of
hearing. In Figure 5.1, I have chosen a note of 200 Hz. Nominally the
violin G string note (196 Hz). With this choice of note, the ear is actually
much more sensitive by the 10th harmonic at 2000 Hz, and so overall,
we hear virtually no change in sound level between them. Had I picked
a really low note at say 50 Hz (e.g. from a double bass) then we would
have heard the 10th string harmonic at 500 Hz appearing to be four
times louder than the fundamental! What we think we hear can be quite
misleading.

The figures are saying (a) the sawtooth displacement pattern fixes not
just the frequency of the fundamental, but also all the possible higher
harmonics. Less obvious is (b) that the higher harmonics do not necessarily match any standard musical note, and (c) our perception of their
relative powers is critically dependent on our hearing response. Even
more surprising is that if we only heard a part of the set, of just higher
harmonics, our brain would still process the signals and claim we were
hearing the fundamental! Brain mathematics overrules reality.

The way this summation of different sine shaped waves can add
together is sketched in Figure 5.1b. The figure plots an example of just the
first three frequencies in amplitudes of 1, 1/2, and 1/3, with the waves
alternatively falling and rising at the start point. If we continue the series
and then add them all together it becomes a more sawtooth pattern. This
is exactly what we will produce from the sticky rosined bow pulling the
string sideways, and then it suddenly breaking free with a fast flyback.
Hence violin strings will generate this entire frequency set. What we hear

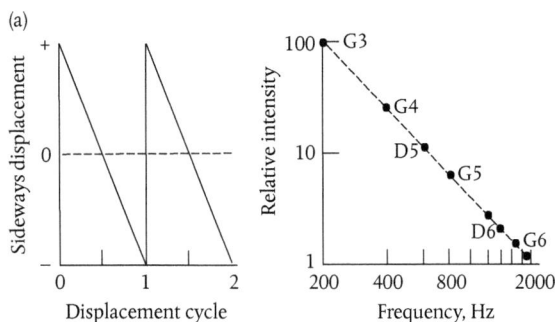

Figure 5.1a The violin bow uses rosined horse hair, so it sticks to the string and slowly pulls it sideways. When the sideways force is greater than the frictional pull, it is suddenly released and it rapidly flies back. This causes a sawtooth movement of the string (sketched on the left of the figure). Mathematically the same type of displacement can be described as the sum of a set of frequencies with a fundamental (f) and higher harmonics that are 2, 3, 4 etc. times this fundamental frequency. So, playing a low string note at 200 Hz (roughly the violin G string frequency) also generates notes at 400, 600, 800, 1000 Hz etc. We respond to intensity of the sound, not amplitude, so the string produces harmonic notes with decreasing intensities of 1, 1/4, 1/9, 1/16 etc. In order to indicate a wide pattern of notes and a large intensity range, the right-hand part of the figure uses both logarithmic power and logarithmic frequency axes. Harmonics which closely match musical notes are indicated. Our ears have different sensitivity at each frequency so the right-hand figure does not necessarily represent what we think we hear.

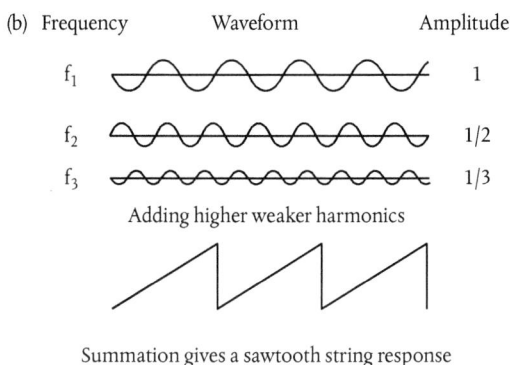

Figure 5.1b The pattern of frequencies for movement of a bowed string which generates a set of harmonics that decrease in amplitude with higher frequencies. The sawtooth frequency is at the expected fundamental of the string.

is then altered by the acoustic response of the instrument. Even from these three, a sawtooth type pattern is emerging and a complete set will remove the ripples on the summation.

The Role of the Violin Structure

An isolated violin string makes very little noise so, as with all musical instruments, we need some form of amplifier. Stradivari and his fellow violin makers were pre-electronics, so the style of amplifier used the vibrations of some wooden plates, and a box with holes for the air to move in and out. Such amplifiers have a lot of resonances, and in a violin, they are chosen to give quite strong resonances near the notes of the three lower strings (G, D, A). This means that these notes are stronger than other notes in the same range, or at higher frequencies. The resonances differ for every note, and of course for the harmonics. Hence, the components of the original intensity pattern of $1:1/4:1/9:1/16$ etc. are all amplified by very different amounts, as the original intensity pattern from the string is modified by the amplifier response of the instrument. The combination of string and violin amplifier characteristics gives the formant for the different notes which we recognize as the sound of a particular violin. I will return to this in a later chapter, where we link instrumental sounds with advances in manufacturing methods and design.

Once again, we are going to be impressed, or surprised, by the brain. It takes the intensity pattern, and does a Fourier transform in a few milliseconds, from which it decides which is the fundamental note. From the pattern of harmonic intensities, it also recognizes the particular instrument that is being played. Literally the specific violin, not just that it is a violin. Identical analyses are made with every sound that we hear. The brain does this in milliseconds, but it is worth noting that although the mathematics for the Fourier transform has existed for 200 years, the time it would have taken to analyse a simple sawtooth string movement would have been measured in terms of many hours until perhaps 50 years ago.

The brain is extremely good at the transformation analysis of any note which has a harmonic content based on multiples of the fundamental. Even if we filter out some of the harmonics we can still quote the base frequency. This is particularly valuable in a situation where the fundamental is missing. For example, a radio broadcast may not contain the bottom notes from a double bass, as radio companies do not transmit the lowest frequencies. Nevertheless, we hear most of the series, and conclude that the player must have been producing the bottom note of that series.

Musically we definitely 'hear' what we expect, not what we detect. Just as socially, we often hear only the content that we wish to hear.

Brain Tricks to Guess at a Fundamental

In the case of the violin string I have a very nice and simple situation where all the higher harmonics are multiples of the fundamental frequency (f). The brain does not merely listen to the set of notes, but it also considers frequencies that appear from the differences and sums of the components. For our simple harmonic series this means all the possibilities such as 4f–3f, 6f–5f etc., and all say that there is a component f (i.e. the fundamental). This just confirms that we have claimed we heard the correct note (plus its harmonics). It is also very useful when some of the

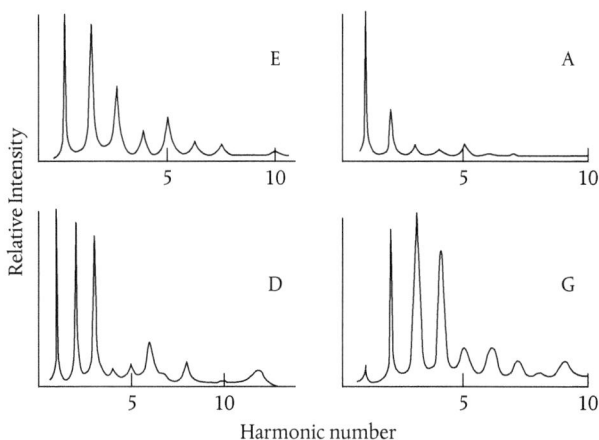

Figure 5.2 An example of the relative power delivered in the main harmonics from a bowed violin. The instrument used here was not of high quality, as is apparent from the response of the G string which has virtually no fundamental power. Frequency differences between the 2nd and 3rd, and 3rd and 4th harmonics etc. all produce the fundamental G frequency, and this reinforces our imagination that it was present. The left-hand signal in each case was the note that was being bowed (i.e. G, D, A, E at 196, 294, 440, and 660 Hz).

components notes have been filtered out, or not transmitted to us. Double bass fundamentals may not survive poor broadcast or speaker systems, but from the higher components we imagine them.

I have one violin that sounds as though I am playing the note G of the lowest string. In reality this is imagination, as when we measured the power level I only detected the higher harmonics. Neither I, nor my students, had realized we were making a mental guess. Figure 5.2 shows the pattern of harmonic intensities that we measured. The A-string pattern is roughly what we expect from Figure 5.2, with a strong fundamental followed by harmonic intensities dropping down to 1/4, 1/9 etc. But the rest are dominated by resonances of the violin. The big surprise was for the G string, where there is virtually no fundamental power, yet we had no difficulty claiming that we could hear it. The strong harmonics of 2, 3, and 4, all offer differences that correspond to the fundamental G. This reinforces our mental prediction of the note.

Anharmonic Signals

In a sense I am making a mistake in discussing the idealized set of evenly spaced harmonics. This is because I am a physicist who plays the violin, and from the science of the strings, evenly spaced harmonics are predicted. However, and this is a very big *however*, many musical instruments do not have higher frequencies that are exact multiples of the lowest note! With high precision measurement, anharmonic patterns appear in everything from metal bars to pianos and wind instruments. Therefore, it is better to find a different word, and to talk in terms of the *partials* or *overtones* of the series, rather than harmonics. If we want to process these much wider ranges of sound tones then we have a problem because it will consume far more brain capacity, power, and energy. So instead we compromise and make intelligent guesses to define the sound from the set of partials in terms of one fundamental note, and ignore the fact that we do not have a simple 'perfect' set of harmonics.

Anharmonic content occurs in the spectrum of frequencies emitted by a piano 'note'. Unlike the violin string, the piano strings are much heavier, stiffer, and more substantial. This gives plenty of power, but the vibrations of the string deviate slightly from the $1:2:3:4$ etc. pattern for the partials of a thin violin string. The heavyweight piano strings have natural resonances that might be better considered as being from very lightweight rods.

Wind instruments all have some degree of anharmonicity, which varies between notes, and the energy level that is blown into them. Musically, this is a great bonus, as the variations are precisely the features that give the instruments tone and character, and allow them to vary in different registers, or between soft and loud passages. It does pose a challenge for the signal processing of the brain, as not only is there no longer a single unique label that we can use to define the note, but the tone quality and interpretation may differ between individuals. Just as with human voices, we can identify the performer.

A relatively extreme example of an anharmonic sets of notes is apparent with bells. The situation occurs both with notes sounding from a set of hand bells, as well as the more massive clock and church tower bells. From the shape of a metal bell the sequence of notes produced from each bell is anharmonic (i.e. there is not a fundamental with higher harmonics in a simple frequency ratio of $1:2:3:4:\ldots$ etc.). Changing bell shape alters the mixture of partials. Further, the anharmonicity will vary between sets of bells. In this situation the brain has to make a guess, and we may then claim we heard a bell sounding a note which is a D. However, if we make a detailed measurement of the frequency components, then we invariably find that not only did the bell not sound *any* D note, but it did not sound any of the exact standard harmonics of a D. We had a situation that was not familiar to us, too complex to analyse, and so we settled for an approximation (i.e. a guess).

As a guideline, the English design of hand bells may include a nominal fundamental note, plus a strong harmonic near the 12th (i.e. an octave plus fifth). By contrast, Dutch hand bells often vary between a strong harmonic on the major, or minor, 10th (octave plus third).

This inability to fully process a non-harmonic series of notes leads to an interesting side effect. If we imagine that we have a source which produces notes with frequencies in the ratios of say $1:2.5:3.5:6.1:7$ etc. *some* brains will process the signals and guess at a series with multiples of a fundamental of frequency 1. The extra checking process will spot the 3.5–2.5 is also 1, and the $7:1$ all look like part of the same series. Other people will hear the same sounds, but prefer to identify with a fundamental at 3.5, as there is a clear note that is double that frequency at 7. Hence two people can often identify with quite different notes if the partials are in an anharmonic set. They are equally justified in doing so, but will claim the bell is of a different note.

Bells and gongs have had a musical history in all cultures. These in turn have led to large-size bells, as for carillons, churches, and big clock towers. They may not all be considered as musical instruments, but they have been the result of considerable technological development over the last few thousand years. Of the many possible metals for their construction there has been a trend to make them out of bronze, which is primarily an alloy of copper and tin (roughly in a ratio of 80 to 20 per cent). This mixture gives both mechanical strength and a good resonance, and critically, a melting temperature that makes fabrication possible. Traces of nickel, lead, phosphorus, and antimony are added, or in Russia, some silver. The famous clock bell in Westminster Palace (Big Ben) started out at 16 tons in Stockton-on-Tees but split during testing. It was then recast in the Whitechapel Bell Foundry at a smaller 13.5 tons, again with a problem, as the hammer used was too powerful, and it caused a split. This was repaired without recasting. The Whitechapel Foundry also made the 1752 Liberty Bell for Philadelphia. Here there was a similar saga of cracks and a split.

Bell-making of large examples is difficult, and the extreme case was the Great Bell of Moscow that was cast in 1733. This was certainly the largest ever successfully attempted. It weighed in at 192 tons, with a base diameter of 23 feet. The bell was fine, but the tower burnt in a fire, and the bell was split when the tower collapsed.

The Strain on the Brain

A simplistic view might be that more complex sound patterns require more processing power. Simple signals (e.g. from violins) fit the basic ideal note pattern whereas other instruments are more stressful as there are decisions of compromises, rather than easy exact assignments of the notes. If we extend this concept to listening to different types of music then the possibility emerges that more classical style composers and/or music that comes in the easy listening category will be the simplest to process. Less stress then means we are more relaxed. Highly atonal music, or with erratic rhythmic patterns, is more challenging and hence more stressful.

This may seem an extreme comment, but there have been studies with animals where there are quantifiable differences in their behaviour depending on the background music that is offered to them. For example; milk production from cows who were milked while listening to soothing music (Beethoven's Pastoral Symphony would seem appropriate) produce around 3 per cent more milk than if they have no music or something more atonal. Fast music slightly reduced their output.

It is known that hens lay fewer eggs if stressed, so music is indeed an economic bonus (Haydn symphony #83, The Hen, or the Beethoven Egmont overture spring to mind). Responses of dogs to various types of music have similarly given clear-cut examples of changes in behaviour. Calm classical music produces more calm and relaxed canines (avoid Offenbach).

Humans show all these standard types of animal behaviour and we are clearly equally responsive to the background sounds, and our reactions and mental state depend on the various types of noise or music. Calm music helps many young people with homework, or undermines the work with the wrong musical choice. Films and TV exploit background sound continuously for mood-setting. Mental conditioning is deliberately introduced for military, religious, or relaxation purposes. While some results are very positive, the more worrying fact is that the animal and human examples suggest that with continuous background music there can be negative reactions. Background noise, or some types of musical sound, significantly raises the stress levels for many people, and this impairs the way they work and function. Recent claims of chemically detectable changes in the body, that are linked to the type of music, do not seem unreasonable, as it is clear that different genres of music need different amounts of brain processing power and stressful decisions. The brain then defines chemical processes which range from pleasant (serotonin) to a range of stress- and tension-generating responses.

I will return to this topic in a later chapter.

A Final Comment on Hearing

The evolution of the biological engineering of the sound detector and signal processing by the brain has resulted in a superb and compact system. Without it we could not have developed speech or listened to music. Modern texts recognize this and instead of using titles such as acoustics, sound, or hearing, they call the topic psychoacoustics. The examples

I have offered in this chapter have hinted at the way we process sound signals and I included examples where we basically made a guess at situations we could not analyse (e.g. anharmonic bell notes). The design compromises equally produced a few oddities which, when we understand how they can occur, might make us slightly more tolerant of singers who seem (to us the audience) not to be totally in tune. However, why we enjoy and appreciate music is clearly in the 'psycho' part of the signal processing, and my view is that we should take the pleasure from it and not worry too much about the science. I am a physicist and realize that we frequently use processes in technology long before we understand the details. This is pragmatism. If we had waited for understanding of mineralogy and fracture mechanics, we would never have reached the Stone Age. Only now do we understand details of the photographic process but have used it for nearly 200 years. Our detailed understanding of the chemistry has appeared at the point in time where the photographic process is mostly superseded by electronic detectors.

CHAPTER 6

SCALES—IDEALISM OR COMPROMISE?

Musical Scales—the Conflict Between Idealism and Compromise

Music has been part of life for as long as we can find records, and while we have no idea of the sound of early singing or instruments we can make some intelligent guesses. The range and tone of human voices might have altered slightly with time, because the average size and shape of modern humans is slightly different from our Neolithic ancestors, but over the past 2000 years the size changes are mostly linked to diet and climate, and certainly well within the current spread of heights around the world today. Thus, our frequency range for speech and singing is unlikely to have changed greatly. What is more obviously very variable is that when we sing a tune, we automatically do so with intervals and scales that are familiar to the region where we are living. The jumps between notes, and the sound intervals that we like to hear, when played together, are also very specific to both region and culture. Only a few intervals, like the octave (e.g. C to C) or fifth (e.g. C to G) are common and acceptable in most cultures. All else is diversity.

In current Western music we expect to use eight notes for an octave (on a piano keyboard the scale of C is just the white keys starting from C). Figure 6.1 sketches a labelled section of a keyboard. The notes of a scale are not all equally spaced and are a mixture of intervals of tones and semitones. Any octave spans a total of 12 semitones (that includes both black and white keys on the piano). The Western C scale on the piano was easy to visualize as it was only the white notes. There are also notes termed sharps (or flats) that require the black keys. For a Western scale, the spacings run as tone, tone, semitone, tone, tone, tone, semitone. So

Figure 6.1 As a guide to basic scales, the piano keyboard is useful. The black notes are called sharps or flats, so D sharp (♯) is the same piano key as E flat (♭). An interval from say C to C is termed an octave, and the higher note of an octave is at double the frequency of the lower one.

even if we start on a black key, say D sharp (also the same as E flat), we can work out the notes of that key.

Rather than an eight note scale, many parts of the world use variants of a five note scale, called a pentatonic scale. To hear an approximate pentatonic scale, an easy option is to play one on the piano, using just the black notes. Start on the F sharp. While pentatonic scales are still in use in many cultures, it is very misleading to assume that the Western pentatonic is the same as any of the others currently, or historically, in use around the world. Examples of pentatonic note spacing, as well as alternative current eight and twelve note (or other!) scales can be heard via web sites, and typically more than 50 examples are on offer. Some may sound familiar, while others seem quite exotic. In terms of total users for a particular scale system, our classical Western scale may not even be the most popular. We may claim it as the peak of our musical development, but in reality it is just one of many alternatives. Indeed, I will deliberately point out that in the piano version it is seriously flawed.

Diversity between scales is not surprising as we are familiar with diversity in language. Although I am writing in one of the three most used languages, my version of English differs considerably from that in other regions of the UK, and even more so from the 'English' of other countries. In some areas, such as India or the Congo, there may be as many as 200 distinct languages within the same nation, and these are undoubtedly matched by variations both in the style of their music and the scales they employ.

Culturally there have been many attempts to quantify the details and justify why we have considered and developed the scales that we use. Any serious study will very rapidly reveal that even in the Western style culture we use more notes than the 12 we can play on a piano keyboard, and within our geographic limits, we have notes in use on say Scottish bagpipes, gypsy music, Hungarian, Spanish folk music, plus many and various types of jazz from the USA, none of which can be represented on a standard piano tuning. Careful listening will also disclose that high grade performers on standard string instruments, as well as classical singers, do not use precisely the same notes in ascending, as in descending, passages. Further, they may adjust the tuning if they are playing with different combinations of instruments (such as a piano or string or wind). Such subtle adjustments in pitch are not possible for a keyboard.

If we are trying to predict future technological changes, then we need to note that modern keyboards on electronic pianos are linked to computer control systems that have the potential processing power to resolve all these adjustments, so that the keyboard tuning could be made to suit different musical scales. It is not beyond imagination to write programs to make the fine adjustments needed for ascending and descending passages. However, these facilities are not yet an option; but initial attempts exist on the latest instruments, so it is a reasonable guess that they will be more widely exploited by future composers and musicians.

Despite the reality of flexible scales that we can hear and recognize from top-flight performers, there is often a perverse fascination to discuss the development of musical scale as though there were some absolute holy grail of a true and perfect example. In view of the 50 or so scales used elsewhere in the world, this is a naïve and very blinkered view. Nevertheless, we perpetuate this error via both musicologists and school textbooks on sound. This is understandable, as musicologists will rarely be trained to truly appreciate the science, and school texts are always highly simplified. Discussions invariably start with the efforts of Pythagoras, and then follow the major pattern of changes in Western music from there on. I will do the same, not least because it will quickly lead to the current role, and influence, of keyboard instruments on our instrument tuning. More importantly I want to emphasize that I believe the driving force behind such discussion was not music but an inherent desire to find an idealized mythical pattern.

Pythagoras and Simple Musical Intervals

Pythagoras was an able mathematician in the 5th century BC, and his focus included a simplistic mathematical investigation of the relative properties of different notes. Music was used extensively in both the

Greek theatre and in the home life of his time. Musical entertainers would enliven home parties (albeit often with separation of the men and the women). So overall there would have been much interest in his studies, especially for the mystical significance which he drew from his observations. Most texts give credit to Pythagoras as the first student of such analyses, and perhaps his was the first good experimentation. However, there are earlier texts showing that similar work was being considered by the Chaldeans in the sixth century BC, and Pythagoras was a student in Chaldea, so there is no reason why such studies might not have been attempted before his time. He has the credit because he was successful in many fields, and so any statement from him is treated with great respect, and remembered.

This may seem an odd comment to make, but musical inventions and innovations can of course happen in different places at different times, and human nature always assumes our more immediate famous ancestors were the leaders. Equivalent reinventions are very common in science and, to stay with Pythagoras, his mathematical rule relating the sum of the squares of the lengths of the sides adjacent to a right angle (90°) triangle, to the square of the length of the hypotenuse is known to most children. He was not, however, the first to realize this relationship between the lengths of the sides. The pattern was discovered engraved in hieroglyphics by the builder of one of the Thebes pyramids. A simple variant of this is to make a triangle with rods of lengths 3, 4, and 5 units as this defines a right-angled triangle ($3^2 + 4^2 = 5^2$). These construction techniques, and the inscription, predate Pythagoras by more than a thousand years. Interestingly Pythagoras was also educated in Egypt, so perhaps he may have had some hints as to the earlier mathematics.

Greek theatre also had plays which we believe were accompanied by music. There was a deep interest in music, and inevitably this means recognizing which notes are harmonious when played together, and which intervals between notes sound dissonant. Pythagoras experimented with notes produced from a simple string to see how the length of the string altered the note, and which notes were in harmony. Musicologists assume that he used a string stretched across two bridges (which in older school texts was called a sonometer). Effectively it is exactly the arrangement for a string on a lute, violin, or guitar. There is more than one experiment that he could have made, and the usual suggestion is that he had a moveable bridge which he could slide under the string (Figure 6.2a). For certain

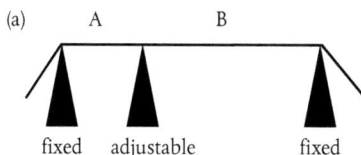

Figure 6.2a A possible experiment of Pythagoras with a stretched wire to find related notes in a scale by sliding a moveable bridge. Pleasant sounds can occur when both parts (A and B) resonate together and the lengths of A and B have a ratio which is a simple integer. The weakness of this arrangement is that every pair of notes would be in a different scale.

lengths, the parts A and B of the system might simultaneously resonate to produce two notes sounding together that were harmonious. The ratios of the two segment lengths were then 1:1, 1:2, 2:3, and 3:4. The simplicity of the numbers encouraged him to think that there was some magical perfection in the relationship between the notes he heard. We should also remember that at that time he was probably only familiar with a pentatonic scale.

As a physicist I would have designed a more accurate, and easier, version of his experiment, which would be to compare notes (i.e. my guess is that the older musicologists had no experimental experience, so I may just be repeating the actual demonstration). Pythagoras could have done this by using a *pair* of identical strings of the same length, and under the same tension, to have two identical notes at the same pitch. This would have meant that he could move a sliding bridge under one of them and, because he then had total control, he would have heard both harmony and dissonance effects for various lengths. The sketch of Figure 6.2b explains this version, where we could look for the favourable ratios of A to B. For a string which is half the length, we hear the same note but an octave higher so B/A is 1/2 (e.g. the note C and the C one octave higher). If the ratio were the modern interval of a fifth, the length ratio would be 2/3 (e.g. C going up to G); a fourth (C to F) would be 3/4, and he may have produced a second (C to D) with a ratio 8/9. He would only have been able to work with relative lengths of the strings, and probably his measurement precision would not have detected anything except the simple prime numbers—plus the fact that he was clearly *looking* for such simplicity. For us, the description is easier to understand in terms of frequency, as this will apply to all musical instruments (not just simple strings). This was not feasible as he had no instrumentation that could record frequency (vibrations per second).

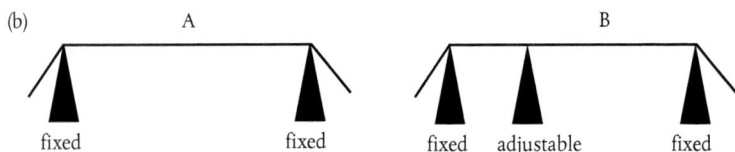

Figure 6.2b A variant experiment where notes from two identical strings are compared. Length A is fixed and notes are played together with notes from section B on the second string as the adjustable bridge is moved to change the length. Experimentally this would have been simpler than that of Figure 6.2a.

In general, string lengths are not useful units to discuss for different notes, and instead we prefer to talk in terms of the frequency of a note in hertz (Hz, i.e. the number of times per second that the sound pressure oscillates). A familiar example is the violin A string, which is tuned to play a note of 440 Hz. It is the same frequency as is used to tune most orchestras nowadays. (In the past, the A tuning has varied with time and country.) The frequency unit of hertz is easy to remember, as Heinrich Hertz was a German physicist who studied waves, and *Herz* is the German for heart, so we can think of the frequency of heart beats.

The notes produced by a string depend on the type of string, and the properties of its elasticity, thickness, and density, as well as the tension applied to it. Our intuition is good in this case as we expect that longer strings, lower tension, and more massive strings, will all cause the notes to be at a lower pitch.

These are now easily measured factors, and we can tie the features together in an equation where the lowest possible note from a string (the fundamental at frequency f) depends both inversely on the string length (L), and on the square root of the tension in the string (T), divided by the string mass per unit length (m). For our generation we write an equation such as $f = 1/(2L) \times (T/m)^{\frac{1}{2}}$. This just states the obvious fact that a long thick string at low tension produces lower notes. 'Our generation' is a slight exaggeration as the expression was derived either by Galileo or the Minorite Friar, Marin Mersenne, in Paris in ~1636.

I am sure that I am typical of a scientist who enjoys music in that I use, and think, in terms of frequency for the science, but only think in terms of the names of the notes when playing or listening to music. So, I will quote both, as appropriate.

Pleasant Musical Intervals According to Pythagoras

To link the pleasant intervals from the sonometer type experiments to frequency and musical notes we can summarize them as in Table 6.1, linking frequency, relative string length, and modern naming of the notes. (N.B. C_4 is the 'middle C' which is at the centre of the piano keyboard and C_5 is an octave higher.)

These four notes are linked by very low number ratios of frequency (or string length). It is doubtful that Pythagoras would have had the precision to measure string lengths for other notes and certainly none would have emerged with such simplistic ratios, as Table 6.2. To an ancient Greek, the simplicity implied it must have a divine origin, not least because the sounds of two singers with such notes would be very acceptable. Idealism had entered musical scales.

For an idealized pentatonic scale (based on the note C) we need to add the notes D and A. The ratio for D is 9/8 (i.e. still a fairly small number ratio) but A is clearly a problem as it comes up at 27/16. We could calculate this by going from C to D (i.e. just one tone) and then jumping a fifth from D to A, so relative to C the C/A frequency ratio is $9/8 \times 3/2 = 27/16$. Pythagoras would have been very unhappy with this.

One may also consider whether, in Greece at the time of Pythagoras, the 'natural scale' and their intervals are precisely the same as in a modern pentatonic Western scale. One should further remember that he trained in Egypt, and at least one pentatonic Egyptian scale differs from the notes in the pentatonic that we might have predicted. For example, in a major pentatonic scale termed Heptatonic, using notes C, D, E, G, A, C, the frequencies are in ratios of 24:27:**30**:36:**40**:48, whereas in one Egyptian

Table 6.1 Frequency and string length ratios

Note	Ratio	Hz	Relative string length
C_5	2	521.4	1/2
G_4	3/2	391.05	2/3
F_4	4/3	347.6	3/4
C_4	1	260.7	1

Table 6.2 Simple wavelength and frequency ratios in a Western pentatonic scale

Musical note	C	D	F	G	A	C
Relative wavelength	1	8/9	3/4	2/3	**16/27**	½
Relative frequency	1	9/8	4/3	3/2	**27/16**	2

variant they are 24:27:**32**:36:**42**:48. Such subtlety would be missed if he were actively looking for idealistically simple ratios and limited by data from a single sonometer, which would have produced pairs of notes in different scales.

Wavelength and Frequency

For music, frequency is the most useful number, and I have already mentioned that speed equals wavelength times frequency, with the analogy that speed of walking is one complete pace (i.e. left plus right) times the frequency of the pace. Interestingly, the Roman army used a complete left/right unit to define a pace when marching. This resulted in our modern definition of a mile as one thousand (mille) complete paces. However, the Roman 1000 paces only came to 1620 yards instead of our current 1760 yards per mile. Try marching with full military kit, and the reason for the shorter stride length will be obvious.

Later Ideas on Scale Tuning

Moving forward 1500 years from the Greeks, our Western music and singing had advanced to more notes than five in the scale, and so several difficulties emerged. Firstly, the simple whole number ratio for say the note B works out to be 243/128. Far worse is that semitone values, as calculated from ratios of different starting notes, arrive at different numbers depending on the route of the calculation! As polyphony and harmony developed, several notes were then played at the same time. These chords include intervals of major and minor thirds (e.g. A major chord of C would include C, E, G, C, and the minor version C, E flat, G, C). Unfortunately, the idealism of Pythagoras of having very simple length (and frequency) ratios, totally collapses once we add these extra notes.

Therefore, a new simple ratio pattern was conceived and, with good marketing publicity, it was called the 'just' scale. It too was not perfect, in that changing between different keys (called key modulation) predicted that different notes were required. This introduces a number of problems and implies that if we had unaccompanied singers, or people playing stringed instruments which were not fixed in terms of all their notes (e.g. a lute), then the singers could learn the 'official' intervals between notes and sing along. It just meant that the notes were actually different in each and every key. Changing key merely meant that they switched to a new note set. There is clearly nothing absolute about this, and in no way were people singing in a 'natural' scale. In singing we can do this without any real problems, our voices are very flexible. We can choose, and are able to control the pitch. As already pointed out, the scales we use are just matters of the current culture.

The real dilemma came about with the development of keyboard instruments, because in these mechanical machines the keys are tuned at the beginning, and there was no realistic way of altering them if a composer modulated between the keys. Players of early keyboard instruments with wooden boards, and tuning pegs, were probably not bothered by the idealized 'official' scales because the instruments drifted out of tune at a remarkable rate. Even modern harpsichords are likely to need a quick retune during a concert interval.

Rather than worry about the music, the solution to all these competing scale options was to go for a mathematical one. Here the approach is to say that over the course of an octave the frequency doubles (i.e. true). There are 12 semitones spread over this interval so, rather than favour one key above another, one can say all semitone intervals are identical ratios. If we go from 1 to 2 in 12 equal ratios then the definition for a semitone means that it has a ratio of the twelfth root of 2 ($^{12}\sqrt{2}$), which is ~1.059. The problem is solved mathematically, equality rules, and *every* note (except the octave) is slightly out of tune in any scale that we can consider. This solution is called Equal Temperament tuning.

Using a modern simple electronic hand calculator, this is a trivial 20 second calculation. We therefore should be extremely impressed that the ratio of adjacent frequencies was resolved experimentally in Italy in 1581 by Vincenzo Galilei (the father of the astronomer Galileo), by using different lengths of plucked strings. Even more surprising is that the same precision was achieved, quite independently, in China by Zhu Zaiyu some time between 1580 to 1584, using different lengths of bamboo pipes. Their musical discrimination, and the precision of their measurements is vastly more challenging than one would even consider to be feasible with such simple equipment. In a modern laboratory an experimental attempt would be far more complex.

Summary of Ideas on Defining Scales

At this point I find it very difficult to follow what has happened to the basic notes that I would have intuitively sung, or played on a violin, now that they are being forced to match the piano equal temperament tuning. To offer some examples of the differences between scales, Table 6.3 lists the semitones within an octave, with frequency values as variously defined by Pythagoras, just, or equal temperament (EQ) tunings. There is

Table 6.3 Comparisons between 3 different scales

Note	Pythagoras		Just		EQ, Hz	Intervals in cents
	ratio	Hz	Ratio	Hz		
C_5	?	521.40	2	528	523.22	1200
B_4	243/128	494.92	15/8	495	493.88	1100
♯					466.16	1000
A_4	27/16	440.00	5/3	440	440.00	900
♯					415.31	800
G_4	3/2	391.05	3/2	396	392.00	700
♯					369.99	600
F_4	4/3		4/3	352	349.23	500
E_4	81/64	329.95	5/4	330	329.63	400
♯					311.13	300
D_4	9/8	293.33	9/8	297	293.66	200
♯					277.18	100
C_4	1	260.70	1	264	261.61	0

also the convention of saying that each semitone spans 100 cents, so an octave is 1200 cents, and the precise note could be described in terms of cents. As mentioned, keyboard sharps and flats are identical notes, whereas they were slightly different notes in the earlier scales, or in fact in the current scales used without keyboard accompaniments. 'Slightly' means that we can hear the difference.

To give us a familiar reference point, I have matched the scales at the A_4 value of 440 Hz. A moderate musician can definitely hear frequency differences of say ~3 Hz in this range, and there will be clear differences between the just and equal temperament scales. Even someone who believes they are not musical will detect changes (just as they would in listening to speech), but they might not consciously recognize what has altered.

Does Equal Temperament Tuning Really Work?

No! The concept is fine but there is a far more serious effect to address. Except for a tuning fork, all the other notes that we generate, whether by voice or instrument, produce not just the lowest note that we intended, but also a set of higher note harmonics (or partials). The balance between these harmonics gives tone quality and distinctiveness, so it is really valuable information. For most instruments the frequencies are simple multiples of the lowest frequency (which we call the fundamental). My violin example said that playing an A at 440 Hz produced components at 440, 880, 1320, 1760, 2200, etc. Hz, (i.e. simple multiples of the fundamental frequency). Our difficulty arises because the higher frequencies do not fit into any standard musical scale, even though we recognize the fundamental at 440 Hz. Table 6.4 shows examples of the mismatch between the harmonics and standard notes; by listing the first ten harmonics of a 440 Hz note it attempts to compare them with notes on the equal temperament scale. Even from these few harmonics, numbers 8 and 10 do not match any official musical note.

For a piano, we invariably have chords with several notes. Each of these individual notes will generate their own subset of higher harmonic frequencies, but not only will each set of the top components not fit to musical notes, but overall, the harmonics of the entire chord will clash and be dissonant. This is most obvious if we have a low note bass chord in the left hand, and a higher register right hand part. A partial solution

Table 6.4 Association of harmonics and musical notes

Harmonic	1	2	3	4	6	7	8	9	10
Frequency	440	880	1320	1760	2200	2640	3080	3520	3960
Nearest note	A_4	A_5	E_6	A_6	$C_7\sharp$	E_7	$F_7\sharp$ to G_7	A_7	Sharper than B_7

to this dissonance is to skilfully *detune* the piano away from the equal temperament scale. The pattern is to make the piano notes progressively more flattened heading into the bass, and progressively more sharpened at the upper end. Rather than call this detuning (which sounds rather negative) we call this technique 'stretched tuning'. The degrees of stretch vary with tuner, the particular piano sound, and the instrument maker, as well as the country (e.g. English, French, and Japanese stretched tunings are all likely to differ). A typical pattern is shown in the Figure 6.3.

Although equal temperament was designed for keyboard instruments, it is comforting to know that not every maker thought it was perfect. The English maker, Broadwood, refused to introduce it until 1846.

While we happily accept this distorted tuning as being typical of the piano sound that we expect to hear, there are some irritating features. As a violinist one can play the open E string, which is in tune with the piano. Within the next octave, the violin harmonic of E (which is easy to accurately produce at the midpoint of the string), sounds flat relative to the piano. The violinist then has to actively adjust to the keyboard. The degree of stretch distortion varies among makers and tuners, so the string players need to be alert to the particular piano that is being used. This is less of a problem for singers as their voices encompass a far smaller range than an instrument such as a violin, plus the fact that detuning is minimal in the range for the tenor and mezzo voices. The highest note of coloratura sopranos may well have conflict with piano tuning, but there are even greater challenges in actually singing their highest notes.

In practice the detuning is definitely not as smooth as sketched here, particularly in the bass. The curve is saying that the bottom end of the tuning range is about half a semitone flat, whereas the higher notes are too sharp by slightly less. Surprisingly, most people never realize that this has been done, and accept it as good piano sound. Truly accurate equal temperament tuning over the whole range produces a sound that

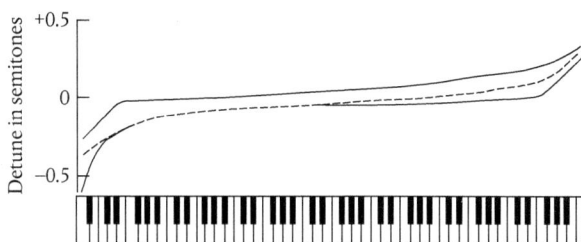

Figure 6.3 This demonstrates the pattern of typical detuning of a piano. The dashed line shows an average target value that flattens the lower notes and raises the higher ones. In reality the detuning is not a smooth function over the range and there will be very large deviations between notes, particularly at the lower end. A typical actual spread is bracketed by the thin trend lines.

we think is discordant. In part this is the problem of thick piano strings, so higher harmonics are raised relative to the harmonics of an idealized thin string.

Naming the Notes

It may seem unexciting to worry about the names we give to the various notes, and I am using the notation where the A at 440 Hz is called A_4. This has the advantage that the lowest note on the piano is A_0. (This is the only piano note with a zero suffix and the next white key is B_1). This is often called the scientific or American notation and indeed, as a scientist, it seems fairly logical. Nevertheless, there are several alternative naming schemes, as devised by organists or musicologists, and just reading a text, without seeing notes on the clefs, can be misleading. Also, the schemes have varied in different periods over the last 100 years. For music books (even the 1991 *Oxford Companion to Musical Instruments* by Anthony Baines) the notation is based on the earlier German labelling of the notes. The pattern described there is to use a mixture of capital and lower case letters with superscripts such as A″, A′, A, a, a′, a″, a‴ etc. This has messy superscripts which are not always clear or easily remembered. For example, the middle C on the piano emerges as c′ in this German notation (Helmholtz). Since both systems are used, I will try to minimize the confusion by offering Tables 6.5 and 6.6. Tables 6.5 just compares two different notations, while Tables 6.6 gives the notes of the strings on

Table 6.5 Notations used by Helmholtz, or 'scientific' or American notations

Notation			Note		
Helmholtz	C	c	c′	c″	c‴
American	C_2	C_3	C_4	C_5	C_6
Frequency, Hz	65.4	130.8	261.6	523.3	1046.5

Table 6.6 String tuning in different notations

Violin	G_3	g	D_3	d′	A_4	a′	E_4	e″
Viola	C_3	c	G_3	g	D_3	d′	A_4	a′
Cello	C_2	C	G_2	G	D_3	d	A_3	a

members of the violin family in two different notations. The benefits of the scientific (also called American) notation is obvious, as there is no ambiguity.

Final Thoughts on Scales

We started with some idealistic mythology that would describe for us a perfect scale. It worked for extremely simple music, but failed with more complex harmonies and key modulations. The mathematical solution, of making every semitone interval an identical frequency ratio, looked tempting and encouraged J.S. Bach to write his 48 exercises to demonstrate the ability to play in every major and minor key. These may be good exercises in keyboard technique, but if the tuning scheme is strictly applied, it makes an unpleasant sounding modern piano. Adding tuning distortions helps the piano sound, but causes clashes with other instruments.

I have not even mentioned attempts to have keyboards with quarter tones or other variants, even though such notes routinely appear in modern jazz music (as well as in avant-garde music, or from singers who cannot stay in tune). Websites show images of quarter tone pianos and they can be found in some museums. When you see them in real life in museums, they look horrendously challenging to play, and (at least to me) sound excruciating. Any detailed discussion of the numerous other scales that are in common usage throughout the world would be prohibitively long. I am finishing with a pragmatic view that here in our early 21st century Western music we can mostly manage with the distorted equal temperament tuning, but I definitely do not think this is the last word in terms of definitions of musical scales.

I also suspect that if electronic keyboard instruments ever match the tone quality of grand pianos, once the manufacturers fully exploit the possibilities of modern computational power, the instruments will offer so much more that they will cause the demise of conventional mechanical pianos. Not least because we will be able to move from the inherent distortions of equal temperament, and choose the tuning scale from alternatives from Europe, and elsewhere, as appropriate for the other instruments that are being played at the same time. Indeed, one can redefine the frequencies of individual notes, and this will allow us to adjust the tuning scale, or add quarter tones, so as to play in any scale from around the world, and/or generate musical sounds that we have not even considered. I see this as a very valuable advance, and it will undoubtedly be exploited by composers. But first, the tone quality of electronic pianos needs to improve. In their defence, the electronic pianos are flexible, and cheaper than high grade concert pianos. Further, the major market for the electronic instruments is probably not classical music, and they are used where tone quality is less important. I may also be underplaying the development of synthesizers, which will compete with my predictions.

CHAPTER 7

MUSICAL CHANGES DRIVEN BY TECHNOLOGY

Why is Technology Relevant to Music?

This may seem an odd question to a musician, not least because, culturally, many will primarily identify with arts and literature, and feel neutral or uninterested in science and technology. Curiously this is not the case for many scientists and technologists, as a high percentage actively participate in and/or enjoy music. More importantly, music has not evolved in isolation from technology, and in general the improvements in instrument design, new instruments, acoustics of concert halls, and latterly electronics, have all had a profound impact on the way music sounds. Inevitably this has been reflected in the style of performance, and these are driving factors for consequent changes in the music that composers write, as well as our appreciation of it. The new opportunities are also allowing many experimental types of composition. With the rapid growth of computational power, I suspect that music produced with computer generated sounds, is still in its infancy.

Understanding some of the technology will therefore improve our appreciation of changing musical styles. In this chapter I plan to give a glimpse of influential features. Equally, it is worth remembering that advances in commercial technologies are relatively transient, and ongoing. Typically, a new idea and product can take say twenty years to come to fruition, be dominant for another twenty, and then fade away as it is totally replaced by new technologies. In part, this matches acceptance by each successive generation, and/or the fact that a major instrumental purchase, such as an expensive piano, may never be replaced by a family. The same is broadly true of musical tastes. Such impermanence is not limited to technology, but also to musical styles.

A prime example of new musically related technologies that emerged, survived for a few decades or so, and then were steadily replaced, is the history of music recording. By the 20th century, electronics had allowed routes to mass production of recordings. They started with early shellac records (78 rpm), moving to 45 rpm, thence into the era of vinyl records (33⅓ rpm). In their turn, they were then displaced by magnetic tapes, then by CDs. CD sales peaked around the year 2000, and then their sales fell with the advent of different format systems, such as MP3 or MP4. The current sales pattern is undergoing a major revision that uses electronic downloads and streamlined access to music stores and this is currently the major market.

The changes in electronic formats and storage systems will equally impact on access to earlier recordings, and our libraries of old examples could easily become unplayable, unless we have kept the playing equipment (e.g. vinyl, reel-to-reel, or other tape decks). This very general pattern of birth, fruition, and decay is discussed in a little more detail in a later chapter. The only consistent pattern is that each technology survives, and dominates, for around 20 years. The sounds from older recordings reveal surprisingly large changes in performance styles with time, and can differ greatly compared with those of today, even for the same compositions and the same performers.

For this chapter it is convenient to artificially split the musical changes influenced by engineering etc. into a number of categories. The first of these includes changes that are not immediately visually obvious but are crucial for modern music. Here I have an easy example with the violin family. The instruments *look* very similar to those from about 1600, and instruments of that period appear to be in use today. Instruments from even 50 years earlier, for example from the makers in Brescia (e.g. Gasparo da Salo or G.P. Maggini), or from Cremona (e.g. Andrea Amati), differ slightly in appearance from a modern violin, but the variations are probably not obvious to a non-instrumentalist. In reality, there have been, and there are ongoing, changes to the instrument designs which are, musically, both critical and obvious. Further, several developments and experimental new designs are appearing that offer musical benefits, so I assume that they will inevitably be incorporated as routine in future instruments.

The second category of instrumental changes is the introduction of totally new instruments, such as the keyed—and then the valved—trumpets, or the saxophone. Effectively most modern instruments have emerged as the result of major improvements in the design of earlier ones. The additions of keys to a trumpet made it more flexible, and enabled it to play chromatically and in any musical key (i.e. to play any of the 12 semitones, not just a few notes like a hunting horn). To some extent the production of the clarinet and saxophone relied on ideas from both earlier and existing instruments. For the saxophone, there was additionally a large innovative step, rather than merely a steady development. At every stage, all these examples offered new tones or more power. This meant that they attracted composers and players who wished to exploit them. Instrument development is continuous, and it is symbiotic with compositional evolution. It is worth remembering that music spans many types, with different audiences and/or special performance situations (military, classical, pop, jazz, etc.). Instruments that are firmly entrenched in say a brass band may rarely be used in a jazz band or an orchestra. So, asking which instruments are musically important generates different answers depending quite strongly on the type of music that we like or need.

My third category of technological input, discussed in the following chapter, is to look at the improvements and changes in design of keyboard instruments. Here the effects range from the way the strings are sounded (plucked or struck) to the continuous developments of the structure, framework, strings, and power. Particularly with keyboard music we have emotional difficulties accepting that a vast number of our favourite piano pieces could never have sounded, when written, in the way they are played today.

Equally important is that we cannot ignore the role of improved concert halls, electronics, and the various ways we have made recordings, plus the influences of broadcasting, film, and DVD etc. All of these areas modify our appreciation of music and dominate our choice of performer and our acceptance, or not, of new musical styles. They directly alter the way composition and performance proceeds, and what the public demand and are willing to pay for.

The Violin Family

When we look at photographs of violins made from about 1600 onwards, there is considerable similarity in their appearance to those produced in this century. With closer inspection some may appear slightly thinner, or with a longer body, while others may have more square shoulders. But these variations exist even between instruments of the same maker. The photographs of old violins are deceptive because, virtually without exception, the violins have been modified for current usage. An original Stradivari violin would have had a short fingerboard, as no composer asked for very high notes. Instead, early violin music frequently required one to play chords, and the bows had an outward convex shape to help play more than one string at once. Further, the strings were made of sheep gut, which did not produce a great deal of power. The description of 'catgut' presumably originated as some players may have produced sounds like the noise of a cat. In fact, there was no fixed pattern of the body shape of a violin, although most makers had a very similar form and often were in a very limited region of northern Italy, and so had apprenticeships and contacts within a small circle of makers. Also, totally novel designs might not have sold.

Probably all of the makers experimented in terms of shape and thickness of the plates etc. as this would be necessary to accommodate the differences between different planks of wood. This meant that there were minor but steady improvements in the design. There were families of violin makers and, for a very long-lived maker such as Stradivari, who died in his nineties, plenty of time to hone their designs and skills. One of his instruments is labelled, 'made in my 91st year'.

A major change was forced on the instrument with the rise of a virtuosic trend, including violinists initially such as Viotti, and then later on, Paganini. They all demanded the ability to play much higher notes at higher speed and with more power. To cope with this the fingerboard was lengthened; the angle of it, relative to the instrument body, was increased; this also slightly increased the overall length of the strings; and the bridge was no longer so flat, so as to allow easier separation of notes on different strings at high speed. Later variations included quite different materials for the strings, in order to increase their power. This use of heavier strings required more tension to tune them to the correct note, hence the body needed reinforcement.

Technology has reduced an early string problem. Violinists today will not remember that one of the features of sheep gut strings was that they were not of uniform thickness and elasticity along their length. Effectively they were slightly tapered. This means that to go to a higher note the spacing to the appropriate finger positions would vary with each string. I have a violinist handbook (Schroeder, 1889) which says, 'be careful to make the strings taper in the same direction'. Failure to do so would mean different corrections were needed on each string, and when playing two notes at once there was a good chance of playing out of tune. Fortunately, modern synthetic strings, with a metallic covering, can be controlled to be uniform. They are also more powerful.

Another contribution to increased string tension, and pressure on the bridge and belly, has emerged because, between the time of Stradivari and the end of the 19th century, the pitch was raised by about a semitone. Hence, the internal bass bar support (which is glued under the low frequency G string) was strengthened, plus other minor adjustments. Many of these upgrades are not visible from outside the instrument. This is a little like car design changes, the external appearance can be similar with time, but the performance is very different. The changing pitch is noticeable for singers. For example, sopranos now need to reach higher notes than they did when music was originally written in the 17th century, whereas bass singers may be pleased that notes in works by Handel may be a semitone easier.

Visually, most of us would not see the differences unless an unmodified and a modern violin were placed side by side. But, in terms of tone and power, there are clear changes between original violins and modified or current ones (using modern strings). Perhaps equally surprising is that the great violins from Cremona etc. have survived the transitions and are still highly desirable. A further positive view is that because the original designs were for both a different pitch and string tension, a modern maker should be able to make even better violins for the music that is played currently. There is a fallacy that makers such as Stradivari had a secret which has been lost. It is true that they had access to excellent closely grained wood from the trees growing in the Italian Alps, which is ideally suited for the acoustics of violins, but there is no secret. The secret is a lot of training and the ability to become superb craftsmen. In the same way, a great artist, such as Caravaggio, was greater than most of his contemporaries because of his skill, not because he had access to better materials. The encouraging news for modern violin makers is that this implies that they can match or surpass the Cremonese instruments, although I doubt the performers will pay equivalent sums to living luthiers.

As mentioned earlier, the violin strings produce very little power, and so the wooden box is intended to be an amplifier. There is amplification by the movement of air trapped inside the body and oscillating through the sound holes. The *f* shape for the hole may seem decorative, but it is engineering, as the curves avoid sharp corner stresses which could cause the wood of the belly to split. It is not a lesson that was always remembered as in the first commercial jet aircraft (the Comet) the designers included square windows, and the stresses that developed at the sharp corners, produced by pressure and temperature changes, contributed to failure of the metallic skin, which resulted in aircraft crashes and deaths.

Similarly, there is an inlay around the edges of the violin back and belly, called purfling, and this is not primarily for ornamentation, but is there to inhibit cracks from running along the grain from the outer edges. Highly skilled craftsmen, including Stradivari, often added more decorative inlays in the purfling groove or across the back, and some makers (e.g. Maggini) used two lines of purfling.

Controlling the frequency of the air resonance is easy; it just depends on the volume of the box and the size of the *f* holes. The real craftsmanship is in the carving of the wooden plates to give amplification over a wide frequency range. This depends on the grain, the density, and the uniformity of the plank. The speed of sound along the grain of the wood is different from the speed in the perpendicular direction. Visually, the violin body is long compared with the width and because of the different sound velocities, the vibrations behave as though it were almost circular, like the skin of a drum. A wave takes the same time to travel back and forth, as up and down (just like a drum skin). As for the drum skin, there are resonance patterns (roughly sketched in Figure 7.1) and a 'good' plate has these tuned to selected frequencies. During carving, the resonances can be heard (just) by tapping the wood, or stimulated by bowing on the edge of the plate, and/or exciting them via a loudspeaker. To make them visible people have used yellow pollen (lycopodium powder) which bounces around, to settle at the nodal regions of minimal movement. Modern technology uses laser holography to achieve the same style of picture. A good plate has several resonances. The skill is to place these at preselected frequencies. Once assembled into the total instrument the objective is to finally have a violin pattern with three main regions of amplification near the frequencies of the D string (from the trapped air in the box) and near the G and A strings from resonances of the plates. Amplification of higher frequencies is needed, but it must not be too strong or the instrument will sound scratchy. With electronics, all this would have seemed easier as it just means designing an amplifier with

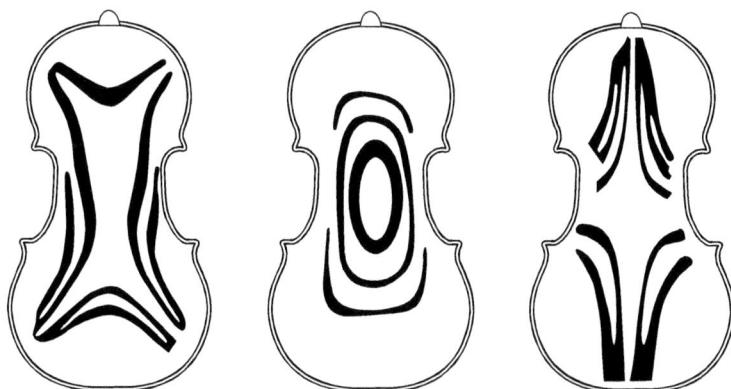

Figure 7.1 Sketches of typical mode patterns of vibrations for violin plates (Waist, Ring, and Cross). The patterns change with excitation frequency. Control of these resonances defines the quality of the assembled instrument.

frequency-dependent gain. Current electronic attempts at doing this produce violins that can be played in the normal fashion but the tones differ so much that for all practical purposes they are new instruments. For classical music their tone is distinctly inadequate, but they are useful for practice without disturbing neighbours.

Just because we now understand the science of the acoustics quite well and have better tools to excite and monitor the vibrations does not guarantee higher quality violins, but it does ensure that we can spot the differences. Commercially it also means we can always make reasonable quality instruments. The subtle differences for excellent violins are not quantified, and still reside with the craftsman. Overall this means that mass produced violin quality has improved from the technology, but superb instruments only emerge from superb craftsmen. (N.B. This now includes some fine examples from craftswomen.)

Violas, Cellos, and Basses

Violas, cellos, and basses are similar in design to the violin but have had less time spent on development. They are still in need of improvements and are obviously benefitting from modern understanding of the acoustics. A standard viola looks similar to a violin but plays one fifth lower (i.e. strings are tuned C, G, D, A instead of G, D, A, E). They should therefore be 50 per cent larger if they were just a scaled-up violin. Physically, the fully scaled-up viola is too big to play under the chin (but possibly acceptable to a musical orangutan). Normal violas sacrifice the ideal size in order to make them more playable, and there is a compromise of a mere 10 per cent increase in body length compared with a violin. This means that the natural instrument resonances do not fall correctly near the string frequencies. That is a major problem because the tone quality is less uniform, and it varies from weak to quite strong as one plays different notes on a scale. Poor tone means that it has been relegated to minor parts in many compositions. It does not stand out well compared with the violins and cellos, with the result that many composers write distinctly unexciting viola parts. The lower pitch also means that it does not naturally fill a role as a popular solo instrument in concertos. It is the poor relation of the violin family, because it has not yet been as carefully engineered as a violin or a cello.

Cellos have fared rather better. Because the instrument is not played under the chin the volume of the body can be increased. This offers help in better positioning of the natural resonances. One problem of too small a cello body is that the resonances are not as evenly spaced as for the violin. In many instruments, two strong resonances sit too close together. This causes an effect called a 'wolf' note when the cellist is quietly playing a note that is close to a natural resonance. Instead of the cellist winning, the instrument jumps back and forth to its preferred natural resonance, giving an unpleasant sound. Improved body sizes for a cello are remarkably easy to construct (while retaining the conventional string length) as one can adjust the body size and extend the lower part (the bouts) of the instrument towards the floor. There is space to do this as it is the region which is normally occupied by the long support spike. Improvements in tone and power (and no wolf notes) are so obvious that I am sure such cello designs will be widely used in the future.

Basses definitely have a size problem. If they were considered to be a scaled-up violin (to give good resonance patterns to match their string frequencies), then the back alone would need to be about six feet high (~1.8 m). This is clearly impossible in terms of suitable wood or playability.

Since the latter part of the last century many people have addressed these problems, and in particular there has been a concerted effort from a group called 'The Catgut Acoustical Society'. They have quantified many of the problems and monitored resonance patterns etc. for all members of the violin family. They have also developed a number of changes in design which have helped to address the tone quality of the lower frequency members of the family. Some ideas still seem unusual and so are not yet routinely used. Increasing the size of an *f* hole lowers the air resonance of the box. Larger *f* holes on the belly are not practical, but there is no reason why additional holes cannot be used in the rib structure, and these are effective in lowering the body resonance. Paying attention to the plate resonances and rib height in the design of violas is also highly successful as it leads to a more uniform spacing of the main resonances, plus a greatly improved viola tone and power. I have two such instruments that were made using these new design criteria. Both have extremely good tone, and because they have retained the normal viola string length, they are easily playable by a violist. For both viola players and composers these are changes that offer totally new perspectives and opportunities for the viola sound.

The double bass has received similar attention, but my personal view on how we might redesign a more powerful double bass (as judged by our ears) is to try a totally different strategy. Rather than thinking in terms of a larger instrument I would actually consider a smaller one! The reason for this heresy is that we are virtually deaf at the frequencies of the lowest double bass notes. Therefore, trying to improve the power output at the lowest frequencies by better instrument design is a waste of effort since the problem is with our hearing rather than the instrument size of the double bass. It is a string instrument, and when bowed the string generates a set of equally spaced harmonics. My suggestion for improving power is to boost the higher harmonics. This is a better strategy because at least we hear them. If we pick a double bass sound at a very low note near 50 Hz (e.g. the hum frequency of the UK electricity supply) this is heard rather poorly, but we should also consider how well we hear higher harmonics (as shown in Figure 4.5). The improvements in the sensitivity of our ears varies with the power put into the note, but if the double bass is being played quietly, then by the tenth harmonic (at 500 Hz) our ear sensitivity can easily be one thousand times better than for the fundamental! This is spectacular, and it means we will detect it. Even if we fail to detect the fundamental, we will still easily hear many higher harmonics. As I explained earlier, the brain looks for a pattern to decide what is the fundamental frequency. For any note with evenly spaced harmonics every example of differences, such as f_5 minus f_4, produces the fundamental. (This was precisely how the violin G string of Figure 5.2 confused us into thinking it produced the fundamental.) Because our little in-built computer has calculated it, it happily claims that the fundamental must have existed. This strategy for low notes on a double bass would work (even if it is at 50 Hz). My logic is thus to improve the intensity of the higher harmonics and leave the brain to make the trickery that strengthens our imagination so we believe we have heard the lower double bass notes. Whether or not we would prefer the tone quality of the modified instrument is impossible to guess, and undoubtedly it would be highly subjective. I am nevertheless sure that double bass players would welcome a smaller instrument to transport.

Improving Wind Instruments

Before discussing the underlying technology for specific wind instruments, it is worth spending a few lines on the way sounds are produced in such instruments. We use construction options with a mixture of uniform tubing, tapered tubing, and expanding bells at the exit end of a wind instrument. Energy is provided by blowing, and the aim is to have a vibrating source (either our lips or a reed) which is vibrating at precisely the frequency we want. If this frequency matches either a natural resonance frequency of the instrument or a frequency which we define by the fingering then we have a powerful note. We can have control over the natural resonance just by changing the length of the tubing or pipework (e.g. by a slide in a trombone, valves in a trumpet, or opening side holes along the tube). So, in the more advanced examples we can adjust and define what are to be the natural resonances.

The wind instruments have several possible resonances defined by their effective length and design, and with a simple instrument, such as a hunting horn or an early trumpet a skilled player can control lip shape and air pressure to hit one of these natural resonances. The effective length is not quite the physical size because the air pressure patterns extend slightly beyond the tube or bell. Sometimes extra notes may be extracted by partially blocking the exit bell to modify the effective resonance pattern (e.g. as in a French horn). Changing air pressure can drive higher notes, and this is termed 'overblowing'. It offers some higher natural tones.

The acoustic principles which define the possible notes may be simple, but I think they are slightly less obvious than for the violin string, so I will compare them. The harmonics of the thin string were directly related to the length of the string between the fixed ends. As no string motion is possible at the ends, these are the nodes (stationary regions). The note we hear is set by a wave running along the string and reflecting back from the far end. For a string, the fundamental occurs when the wavelength of the vibration matches the return journey; in other words, the string length that we see defines a *half* wavelength, Figure 7.2. Higher notes appear when there are nodes along the length of the string. If we count and think in terms of half wavelengths ($\lambda/2$) then we have $\lambda/2$ times 1, 2, 3, 4, etc. giving the familiar set of harmonics of f, 2f, 3f, 4f, 5f etc. (The velocity is fixed and so, as for walking, if we halve the step size then we have to make twice as many steps per second).

The positions of nodes for a wind instrument are not as tangible as the limits of the string. The nodes will exist where the pressure is not changing. One such place is at the end of an open tube. An open end for a tube in resonance is always at atmospheric pressure, and thus unchanging (i.e. a node). There are therefore different response patterns for a pipe which is open at both ends and one which is only open at one end. The differences are nicely demonstrated by comparing the idealized harmonics of a flute and a clarinet. Both instruments are of a similar length but their tone qualities are very different, and they play at a different pitch. The vibration patterns are compared in Figure 7.3.

The flute is excited by air blowing across an edge near one end, and the other end is just open. This means that both ends are at atmospheric pressure, and so both act as nodes. All the pressure changes and fluctuations

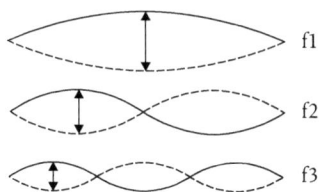

Figure 7.2 The pattern of half wavelengths defined by a string with fixed ends (e.g. the violin family). The velocity of motion along the string is constant, so if we halve the node spacings we double the frequency. Overall, such a bowed string produces harmonics that are 1, 2, 3, etc. times the fundamental frequency.

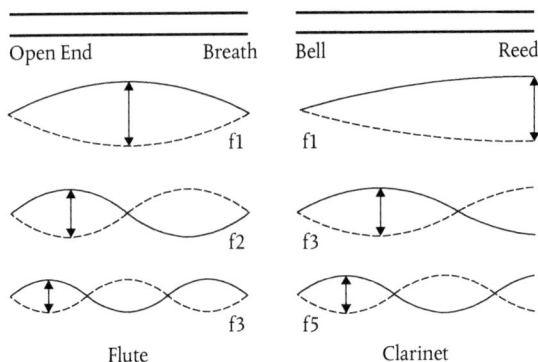

Figure 7.3 A comparison of wavelength patterns for flute and clarinet style tubes. Note that they have a different harmonic mixture above the fundamental. The nodes of zero motion occur where there are no pressure variations. This happens at an open end which is at a constant atmospheric pressure.

for the antinodes are within the tube. The tube length defines the harmonics, which are multiple half wavelengths of the length of the flute. This resembles the violin in that there are a complete set of possible half wavelength alternatives, so there is a full set of harmonics f, 2f, 3f, 4f, etc. Indeed, at high notes the flute and violin can sound somewhat similar. Note, however, that the violin string was forced to make a sawtooth movement because of the frictional sideways pull and a fast flyback (Figures 5.1a and 5.1b). There is no such constriction for the flute, so the harmonic intensities need not follow the same intensity pattern as for the violin string (e.g. Figure 5.2).

By contrast, the clarinet has only one end constantly at atmospheric pressure, at the bell. The energy entering at the vibrating reed end is not at atmospheric pressure, and it is changing. Therefore the body of the clarinet defines alternatives of 1/4, 3/4, 5/4, etc., wavelengths. This means (a) it only has odd harmonics, and (b) the lowest note is one octave lower than a flute of the same length.

Figure 7.4 contrasts two sets of measured harmonic content which show this difference is broadly as we have predicted. In this example, the flute note was G_4 and one sees evidence of all the set of harmonics decreasing quite smoothly in terms of relative intensity. The fall-off in intensity is similar to that of a violin string example for A_4 seen in

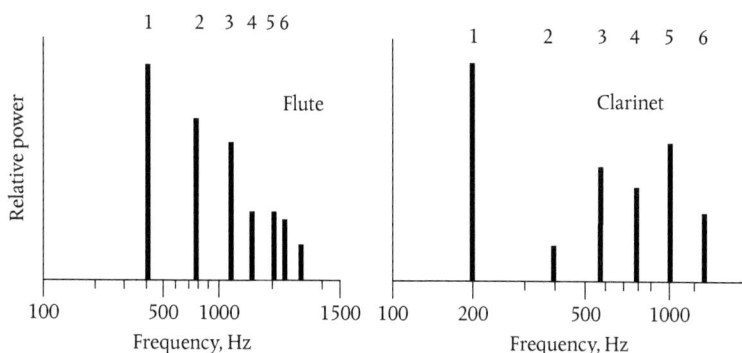

Figure 7.4 This contrasts the patterns of intensities emitted by the sets of harmonic overtones from a flute and a clarinet. The flute typically has a decreasing intensity pattern of every harmonic, whereas for a clarinet the odd harmonics are normally the strong components. Although not shown here, there is often a background in both instruments. The precise responses differ with the performer, intensity, direction, and instrument.

Figure 5.2. In Figure 7.4, the four harmonics that are octaves of the flute G are strongly defined, but other frequencies are also present. The clarinet example (for a B_3 flat) has a mixture of strong odd harmonics (as predicted), plus a much more confused pattern of weaker ones at higher frequencies. There are five B flat harmonics, plus some strong E flat harmonics within the remainder of the pattern. Note that all such analyses are very sensitive to the specific instrument and the player, so further detailed analysis is irrelevant.

In my simple idealized sketch of Figure 7.3 I am still calling these notes harmonics, because for such a simple model they would be exact multiples of the fundamental. This is not the case in most wind instruments because there are bell-shaped endings to produce more power, and also because many wind instruments have some or all of the instrument built with conical tube (e.g. an oboe or bassoon). A conical or any other form of expanding tube-work produces higher notes that are not exact multiples of the fundamental. I therefore should rephrase the discussion in terms of partials or overtones. Indeed, it is these anharmonic characteristics that make the wind instrument tones so interesting. Skilled players can alter the balance between the natural overtone intensities to offer an even greater range of colour.

Examples of Instrument Evolution

The continuous attempts to offer more notes, more power, and greater flexibility in the sounds are matched by changes in the music that is written for them. An early brass example with a small range of notes was the trumpet. As for the hunting horn, this only offered a few notes from a limited range of overtones that could be selected by lip control of the trumpeter. As an orchestral instrument, it was therefore fine for a fanfare or an occasional addition to a work, but definitely it was not ideal for solo performances, although a number of reputable composers wrote short concertos for it. This situation changed significantly with the addition of holes along the length of the tube that could be open or closed depending on pads that covered the holes. These were initially operated by keys, and this first version of a keyed trumpet did not extend the range of the instrument but it did allow access to a complete chromatic scale with every semitone. The difference was of course spectacular, and Haydn led the way in 1796 to popularize the instrument with a keyed trumpet concerto performed by Anton Weidinger. Haydn had a sense of humour and commenced the work using only notes that would have been feasible on a keyless trumpet. The latter part used the new chromatic capability and, of course, was a major success. Hummel followed in 1803 with another concerto for Weidinger and his new keyed instrument. It then became firmly popular, especially with Austrian military bands, and survived for another 20 odd years.

In order to cover a greater range of notes, a set of trumpets were initially needed that separately spanned a relatively small range. Tone qualities were not identical, as each version had a characteristic degree of brilliance and penetration of the sound. Clearly a limited range of notes, albeit chromatic, left room for improvement, and this came in the 1820s via valved trumpets that sent the air flow around different lengths of pipework to create a much wider spread of notes. Normally a combination of three valves was sufficient, but for exceptional works a fourth valve can exist, although the plumbing appears very complex. Additional shank sections can also be inserted to lengthen the total path, and so provide lower notes. The benefits for composers and performers were obvious, with enthusiastic writing from Beethoven to Wagner, and onwards to modern composers, and improvizations from the soloists that offer star quality fame.

Many trumpet variants have different degrees of taper up to the horn, and this in turn means the trumpets are clearly anharmonic so, if used over a wide range, the top fundamental notes will be out of tune. This is not the same as having overtones out of tune, because for overtones, the anharmonicity is part of the desirable and distinctive tone quality. To compensate for sharp high fundamentals, some versions have a short sliding section that can be used to bring the notes back to an acceptable frequency.

The ongoing developments have refined the trumpet design and construction, as well as bringing the instrument an extremely wide following across the musical fields from military to brass band, to symphonic and jazz music. The additions of keys and then valves are an excellent example of technology-driven changes in musical composition coupled with a feedback to make yet further changes that have totally revolutionized the role of trumpets. Basically, equivalent patterns have proceeded with all members of the brass instruments although some have remained in particular musical genres.

The Clarinet Family

As a second example of technological developments with wind instruments, I will pick the variants of the clarinet. Here we have a reed wind instrument designed in wood with a distinctly different sound from any of the brass discussed above. While the family tree has many ancestors, the clearest predecessor was the chalumeau in the version made by J.C. Denner of Nuremberg. This was a wooden reed instrument with various holes to select notes, either by fingers or by two keyed pads. It had similarities with earlier recorders and spanned about one and a half octaves. While the chalumeau went out of fashion, the lower clarinet range is still termed chalumeau. Denner and/or his son Jakob added more keyed structures to ease fingering for a wider range of notes, and by about 1710 instruments called *clarinettes* were being produced.

The tone was acceptable and compatible with other instruments of the period, and there were steady improvements in design. By about 1839 a clarinet marketed by Hyacinthe Klosé had sufficient keys etc. that it now resembles a modern instrument. Nevertheless, in order to cover a wide range of notes the fingering was difficult, and this problem was addressed by Theobald Boehm with a Paris patent in 1844. Interestingly, compared with say violins, the key and fingering mechanisms are still not universal, and although Boehm is widely used, some Germanic clarinets still use alternative structures and techniques, both in the number and function of the keyed notes and in the design of the mouthpiece. However, to a listener they are clearly still all clarinets

As with all the wind instruments the fact that they have a limited range has resulted in a profusion of similar instruments of slightly different size to cover different notes. In particular, the clarinet includes a full chromatic range, but merely by air pressure it is not easy to extend the range, whereas in some other instruments the player uses the technique of overblowing to drive higher overtones. Not only is the clarinet range limited, but any overblown notes are often out of tune and/or one needs even more key combinations. One solution is to use different clarinets for alternative ranges. But this means that on changing clarinets for different needs the performer has to reconsider the complex fingering that would be needed for each instrument. To sidestep this problem the clarinets are called say an E flat, C, or B flat instrument, and the music is transposed so that it appears to be in the key of the instrument. Therefore, the musical score for a C clarinet may actually have the note C written and this implies a specific fingering for this scale. The same music played on an E flat clarinet, with an unchanged fingering, will produce an E flat. To ease the efforts on fingering, the printed version of the music has been transposed, so that the written version produces the correct note on the selected instrument.

To a violinist or classical pianist this seems a little odd, but for a wind player it is helpful and standard practice. For an electronic piano, electronic transposition is available, so accompanying a singer in a different key only requires one score, and an electronic resetting of the pitch. Nevertheless, some pianists still say it is confusing to play notes that sound at a transposed frequency. There is an identical problem for singers who may have a score written in one key, but the performance has been transposed to another.

The Saxophone

The saxophone is unusual in that rather than it being a natural development of an earlier instrument it was a conscious invention to address a weakness in the sound of military bands. At the time, in the early 19th century, the military had the big powerful brass instruments and much lighter weight sounds from woodwinds. Adolphe Sax, from a family of Belgian instrument makers, decided he could smooth this divide with a much larger reed instrument. He therefore invented and developed a very large tapered bore instrument with a clarinet style reed mouthpiece. To make a large powerful sound he required a large taper and horn, and because of the size and shape, it was fabricated in metal. Also, as it was large, it meant that the sound holes along the length also had to be large, in order to force the nodal points that define each note. Clearly there was no way these could be covered by fingers and so he added a padded key system. In 1846 he patented it in Paris under the name of a bass clarinet. He was definitely as skilled at marketing as at instrument making and he ran a competition between two military bands, one of which included his 'bass clarinets'. This was the winner, and so the saxophone then entered into the military bands. Variants of it have stayed with military bands ever since.

The only negative feature was that he was not well received by the classical symphony orchestras of his time, so saxophones still have a minor role in classical music. They have thrived and become essential instruments in military, dance, and brass bands, plus many virtuosic players in jazz. The original design of Sax underwent numerous improvements in order to a cover a wider range of notes, and an entire family of instruments was developed with many competing patented features. There are also alternative fingering schemes, especially to hit the higher notes, and key arrangements to give a linear chromatic fingering (i.e. as for a piano). The considerable popularity of the saxophone family across a wide range of music means that the instrument design, materials, pads, fingering arrangements, and construction are still actively progressing. Musically the tone is very versatile, and my own enthusiasm for the sound probably results from the fact that my father played the saxophone very well.

CHAPTER 8

DEVELOPMENT OF FIXED FREQUENCY KEYBOARDS

Keyboard Instruments

The ancestry of the modern piano includes instruments such as the harp, the clavichord, and the harpsichord, as well as a mass of developments under the umbrella of 'piano'. Unlike the string or wind instruments mentioned above, the piano is not portable. The trade-off is an instrument that can play many notes at one time, and have music that makes it self-sufficient, without the need of other players. This means it has been a crucial item to allow amateurs to play and enjoy music in their own home, either solely as a piano, or in conjunction with singers or instrumentalists. It further has the feature that every note is pre-tuned, so hitting the correct key gives the desired note. The alternative is one of the difficulties for string players and singers, where good intonation requires much practice and accurate hearing.

Starting with a non-keyboard ancestor, the harp, the frame may look like a vertical grand piano but there are fewer strings and to achieve sharps and flats require pedals to alter the string lengths, a system which requires some skill in chromatic passages. The method of exciting the strings is to pluck them. Despite this, some people have described pianos as a harp in a box. A tiny but useful modern advance has been to colour code the harp strings.

In keyboard terms, predecessors included the clavichord, where a metal edge was hit against a string, and a harpsichord where the string is plucked by a plectrum. Technology of the plucking action has moved from plectra made from bird feathers (a quill) to leather and plastics (e.g. Delrin, a polyoxymethylene plastic). Other elements of clavichord and harpsichord construction have similarly been modernized, but basically the current aim is not to alter the sound, but to have a more reliable version of the early instruments, complete with their original musical characteristics. Although instrument development for these instruments has not totally stagnated, as improved materials are added or strings are improved so that they do not rapidly drift out of tune, the musical reality is that there is no obvious desire to change the sound of the clavichord or harpsichord. They are mostly seen as period instruments that play and sound as for music written several centuries ago. In homes there are relatively few examples, and consequently there is a limited impetus to write new work for them. Although this is my instinctive view of the music for these instruments I also recognize that the modern electronic pianos are particularly adept at simulating the sound of both clavichords and harpsichords, and as the electronic keyboards enter further into the home piano market there could easily be a revival, both in the playing of the existing repertoire, and a greater incentive to write more for them. In my experience, many models of electronic pianos are more successful at harpsichord simulation than for the piano. The bonus is that the electronic ones stay in tune, whereas harpsichords perpetually drift out of tune.

Clavichord and Harpsichord

The clavichord is a very quiet instrument, and the action is sketched in Figure 8.1a. The figure shows that striking a key drives a metal edge against a length of wire string. This splits the string in two parts, and so it should produce two notes set by the lengths of the two parts (as defined by the tension in the string etc.). To kill the unwanted note, the shorter string section has felt in contact with the string, so any sound is very rapidly damped out. The sound from the other part continues until the key is released and then the felt deadens the entire string. The problems are therefore that one must keep the key depressed for the duration of the note, and the unwanted section of the string may also initially contribute to the sound.

The plucking action of the harpsichord (Figure 8.1b) is rather different and the excitation is driven via a pivot between the point of contact on the key and the string. The movement sends a plectrum to strike, and go past, the string. The string then continues to vibrate while the key is depressed. When the key is released the falling plectrum will hit the string for a second time, but during this part of the action a damper hits the string and hides most of the second strike and silences the note. Other variants have alternative design actions to avoid a second contact. Again, the key must be held down to sustain the note, but the pivot actions showed that one could gain power in the striking phase. Real volume control is very difficult, but high-speed playing is feasible—if suitably skilled! In reality the struck string intensity fades extremely quickly, so to maintain the interest in the music, most composers of solo harpsichord music offer many notes. Scarlatti's music is a classic example of this approach. By contrast, where the harpsichord is used as a background continuo for choral work (as in opera) there is no need for a persistent sound.

Figure 8.1 shows that for the two keyboards the string directions are arranged quite differently, with the harpsichord in the more familiar direction of the grand piano.

The harpsichord was a significant instrument for a long period of time and among the many design features that were included were the addition of more than one set of strings. The fundamental notes (one or two strings) were often mirrored by a set of half-length strings which added an octave above the lowest note. The two sets of strings are referred as 8 foot and 4 foot, and it was possible to have either the lower frequency or the pair being played together. A second bass string offers more power. For more volume control, shifting the keyboard action sideways selected only two of the strings, so there was some loudness and tone control, but only in a stepwise fashion, without any gradation, as one would have liked.

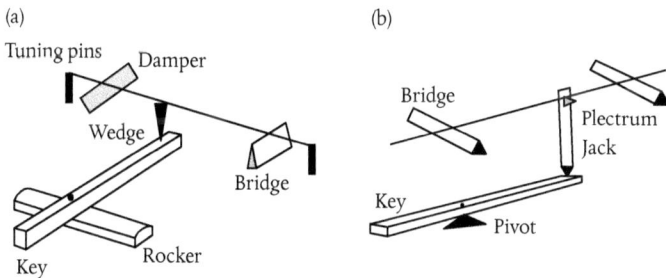

Figure 8.1 A comparison of the tangent striking action of a clavichord (a) and the plucking action of a harpsichord (b). Note that harpsichords can operate with more than one string per note to give more power; also there can be an extra string which is an octave higher. Dampers etc. suppress any secondary plectrum contacts.

Piano Development

By around 1800 the dominance of the harpsichord started to fade as new piano type instruments had been improved to the point that they were preferred. This was not some sudden revolution because the first attempts at using a piano action of hitting strings with some type of hammer began around 100 years earlier by Bartolomeo Cristofori in ~1700 in Florence in Italy. Financially this was not very successful as the Italians were obsessed with opera, and the obvious weakness of a piano is that it is a percussion instrument. There is some control over the initial intensity of the sound (i.e. far louder than for a harpsichord) but the volume fades away and, compared with control with the voice, it is unimpressive. Cristofori was therefore neither rich nor remembered. The piano survived with developments, particularly because of the efforts by Gottfried Silbermann in Germany, where he improved the action. He was inspired and influenced by J.S. Bach, and they worked together. Bach saw the instrument in a different light from the Italians as he was an organist and recognized the advantages of the new machine. His organ background meant that he was less concerned with the fact that if one altered the starting note of a tune (e.g. from say the key of C into E flat) the slightly discordant sounds of some notes were acceptable. I will expand on this later.

From the basic Cristofori, design improvements and variations took hold and there were early contributions from German and French instrument makers. The pianos that we hear today are very different in power and performance, and experimental changes are still being offered, with results that would probably be quite unacceptable to players of the 17th century.

The essential novelty of the initial designs was that, via a lever system, the action of pressing the piano key both raised a damper off of the string, and simultaneously accelerated a padded hammer to hit the string. The hammer bounces, and this allows the sound to continue, until the finger is taken off of the key, which causes the damper system to fall back in place. Overall, the immediate difference from the earlier harpsichord style instruments was a strong sustained sound that is under the control of the pianist. In engineering terms this is more complex than the earlier instruments as it had a complex sequence of actions.

Crucially, the pivots and lever system that propel the hammer against the string deliver a blow which is related to the finger action, so there is some control of volume and attack. In the early Cristofori pianos, the

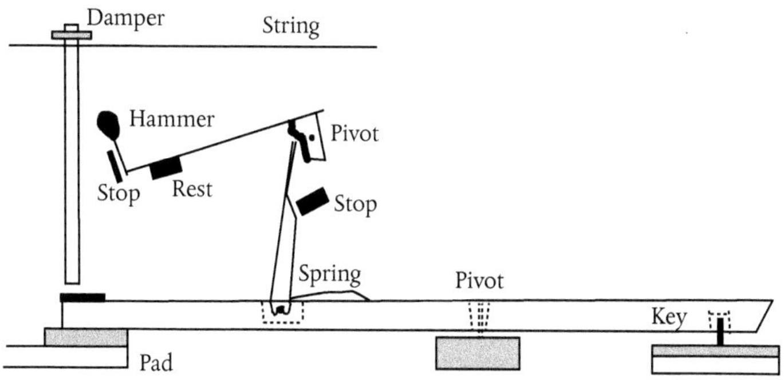

Figure 8.2 A very simplified view of the action of an early English grand piano.

design was a little different, as there were springs and check controls to hold the dampers, the hammer, and the lever mechanism. A *very* simplified sketch of the action is shown in Figure 8.2. Rather than use the Cristofori design I have chosen a much later English model of about 1795 as it is closer to a modern piano action. Even this is simplified, and many details are omitted.

Several advantages and possibilities for improvement became apparent. Because the hammer action bounced off the string, it was possible to have a separate foot pedal mechanism to stop the dampers from returning, and the note could sound while other notes were played. The extra power of the lever system meant one could use heavier weight strings, or even more than one string per note (as in the later harpsichords). This meant more power. Undoubtedly it would not have been predicted, but there is a further bonus that with a pair of (or three) strings per note, there are inevitably some tiny mistuning differences. Since the decay of the intensity for each string is slightly different the power resonates between them and the effect actually slows the speed of the intensity decay. Energy transfer for piano strings is the same process as happens if one pushes one of two children's swings mounted on the same frame. Because the frame flexes, when one swing slows down, the other picks up the energy and moves. Also, because the hammer bounced back from the string, and was independent of it, one could add a more rapid hammer recovery system so that notes could be repeated at high speed. Without this rapidity the virtuoso performances of say Liszt or Chopin would not have been possible.

A second advantage of higher speed, which is never discussed, is that we do not hear each note for very long, so the fact that piano tuning is a compromise is far less obvious.

With such improvements there was an incentive to add more heavy-weight strings to gain yet more power and by say 1800 the piano had reached the version we now call the fortepiano, which would have been familiar to Mozart or Beethoven. This was adequate in terms of volume, not only for a home piano, but also for a small concert hall of the period, in conjunction with an orchestra. Unlike music in a harpsichord concerto, both orchestra and soloist could be heard together (i.e. harpsichords are often swamped when the orchestra is playing). As mentioned earlier, the structures were made in wood, and mechanically this limited the total string tension (both in terms of string mass and the number of strings per note). It was a very much quieter instrument than any modern piano. Nevertheless, the many developments included addition of a wider range of notes, both at the low and high end of the piano. All such features had a direct influence on composition, power, and the virtuoso opportunities in concert performances.

The wooden frame was the main limitation in piano design at that stage, and so the next major advance came during the 19th century with addition of cast iron frames to cope with the force from the strings which can be equivalent to ~20 tons (~20,000 kilograms). The structural strength offered new options in the layout of the strings relative to the sounding boards. Less obvious is that thick strings have a slightly different harmonic response from thin strings, and so the higher partials (overtones) are not exact multiples of the fundamental frequency. The new overtone patterns can either be viewed as distortions (relative to simple physics), or as a bonus, as they modify the overall tone quality.

As I mentioned earlier, violinists find that piano tuning can be irritating at higher notes, which appear beautifully in tune as harmonics on the violin, but which clash audibly with the same 'nominal' note as tuned on the piano.

Dilemmas with Piano Music

There are real conflicts and arguments among both musicians and the listening public on whether we should use ancient designs or modern ones when playing early keyboard compositions. Even from very fashionable composers such as Mozart or Beethoven we need to remember that they were often writing for harpsichords or early versions of the forte-piano. This means that their keyboard music was composed for totally different instruments from those in use today. For them the instruments were changing in range, power, and performance but they would never have been able to imagine the sound and responses of a modern instrument.

There is absolutely no clear answer as to which to use. I have heard performances of the Archduke piano trio by Beethoven in which the modern grand piano was well played but totally overpowered the violin and cello. Every aspect of any of the pianos used by Beethoven differs from a modern machine, from the attack sound of the start-up of the notes, the balance between low and high registers, the persistence after a note is played, to the feel of the action to the pianist. Plus, the definitely non-trivial feature that the maximum power has gone up by nearly a *thousand* times in the last two centuries! This means that we hear it as eight times louder. The power from violins, violas, and cellos has increased slightly, with different types of strings, but nothing by comparison with the piano.

Musically it would seem obvious that we would want to hear my Archduke trio example of 1811, rendered with something closer to a fortepiano, not a modern pianoforte. Unfortunately, we are so conditioned to the modern sound that it is difficult to accept the earlier one. We should also remember that Beethoven was not happy with the sound of pianos available to him. So even with Beethoven the situation is a mess. For his piano concertos, he thought the fortepiano instruments were not powerful enough against the small orchestras of his time. At the time, a 'big' orchestra had fewer than 30 stringed instruments, which is often only half that of a modern orchestra. Orchestras are now much louder, as well as being larger, so for a piano concerto we may need the power of the modern instrument. Musically, we are enjoying something quite different from anything heard at the time of the original composition. With modern conditioning, and exposure to new piano sounds, we probably would not appreciate the original.

These are two scenarios from the same composer, but probably quite different outcomes, and certainly never with universal agreement.

In the case of the piano the changes are physically clear, as we have progressed from lightweight fortepianos with wooden frames, which could only withstand the tension of thin strings, to iron frames and then steel ones. In parallel, the string weight, thickness, number of strings per note, and overall tension have gone up, to produce more power. Physically we see this just by the weight of the instruments. Changes were happening in Beethoven's lifetime, and by 1824 the Conrad Graf piano from Vienna had a wooden frame of ash and spruce, thin metal strings and weighed in at 390 lb (i.e. already more power and heavier than the early fortepianos). By this century a typical big Steinway with metal frame, plus more and thicker strings has pushed the string tension on the frame up to some 20 tons(!) in an instrument weighing around 990 lb. The later Fazioli 691 is even more powerful and is a 1520 lb (691 kg) heavyweight.

All such changes are continuously altering what we hear and associate with piano tone. Equally, we can only conjecture as to what composers would have written, and how they would have performed, if they had had the opportunity to use modern instruments. In the case of the piano there are current developments that are still very definitely active, and instruments of the present time may in the future seem as dated as the fortepiano.

The timescale of each successive piano generation in the past was around fifty years. This will shorten with the advent of more rapid flexibility offered by the electronic variants. Not least is that a feature of the most recent electronic pianos is that they can almost simulate the sound of different earlier vintages. With better electronics, the simulations should improve. Once these new instruments penetrate a wide market, we will routinely be able to reassess music as written for the fortepiano or 19th century instruments that were appropriate to the composers of that period. If this happens, modern composers may even consider writing for such instruments as their music would then be widely playable. It is very clear that electronic piano tone is not yet 'perfect', if we compare it with a high quality mechanical piano. I assume this performance gap will narrow, and the real potential power of the electronic keyboard and its associated computing power is not going to be just an electronic high grade modern piano, but an instrument that will be able to emulate a wide range from fortepianos and harpsichords etc. to more instrumental sounds than they do at present.

I cited the problems of live performance of say the Beethoven Archduke trio with a modern grand piano (plus a violin and cello) and commented that it was unimpressive because it is difficult for the pianist to keep down the volume and attack, so as not to drown the music from the two string players. Mostly people hear broadcast or CD performances and may not feel so critical about the balance because the sound engineers boost the volume of the two string players, and they may also filter some of the piano frequency response.

I have catalogued a large number of changes, and with so many possible and necessary developments it was inevitable that they took a long time to emerge. A view of a modern piano action, whether in a grand or an upright, may have a broad similarity with an early Christofori instrument, but it is now vastly more complex. I have not troubled to add a diagram of the modern action as the design subtleties will be hidden for those of us who are non-expert in the mechanics of the action. However, detailed texts and websites show such actions in considerable depth, and typically they list more than 80 *main* components on the mechanical action of *each* note!

Electronic Pianos

While the keyboard may closely resemble that of a mechanical piano, and in terms of touch and response there have been successful efforts to make the electronic instruments feel very similar to the stringed version in terms of tone and other features, we really should consider the new electronic versions as a completely new family. Indeed, as in most families there are some members that are attractive and show great promise for the future, and there are other members that are far less desirable, or appeal to a totally different audience. This is immediately apparent as the electronic generation of sound is incredibly complex if one wishes to duplicate the sound of existing mechanical instruments.

Matching a harpsichord tone is reasonably straightforward as it is a simple thin string that is being plucked and so the harmonic patterns and decay characteristics are very reproducible. Further, most audiences and home performers will not be totally familiar with the original harpsichord sounds, so most of us will find the tone quality acceptable. Two other major classical keyboard instruments are the piano and the organ. Of these, the organ tones seem most successful over much of the frequency range (at least to a non-organist), but piano tones were initially far less appealing. Some makers have tried to overcome this by having electronic memory copies of high grade piano sounds, which they duplicated in blocks over the spread of the keyboard. This is partially successful, but there are often quite obvious tone changes as notes move from one block to the next. At least one maker has taken a very different approach to try to model every aspect of sound production within a keyboard machine, and then use powerful computing methods to simulate the sounds of every note. This route was impossible 30 years ago with early machines, but the electronics and computing power have advanced to a point where it is now feasible (and moderately successful).

There are also many types of piano sound (e.g. just from modern instruments), and so some players prefer a bright tone, others a mellow one; some prefer a strong bass and others want a sweeter mid-range tone quality. Yet again there may be situations where one wants a 'pub piano' sound etc. For an electronic piano maker these differences were initially severe challenges. To a large extent, the mass market for electronic keyboards may not be for classical music or concert halls but for pop music where it is essential to offer many other instrument sounds and effects and the 'piano' tone is not the primary objective.

The world of semiconductor electronics with computer chips and large memory stores has only existed for half a century, and the development pattern is phenomenal in that both chip speed and memory capacity are continuously increasing. This is to be contrasted with the design and development time for an electronic piano, which is probably on a timescale of a decade or more, from planning to market penetration. Hence every new instrument is out of date by the time it is marketed! A further delay factor is that, once it is purchased, a home user is likely to keep the same machine for perhaps 20 years, and effectively new models are going to a new generation. Overall, this means that one can predict what is immediately feasible, indicate what is desirable in the near future, and offer suggestions for totally new modifications which do not exist on any instrument at the current time. However, the futuristic items may not gain instant mass appeal, and parallel markets of say pop music synthesizers may be financially more attractive than instruments aimed at classical music. Also, unlike the purchase of a high quality mechanical piano, which may even appreciate in value, the electronic version will become dated and fade in value. There are two bonus factors. The first is that one does not need to pay for one or two tunings per year, and the second very major selling point is that the purchase is not of a single piano, but an entire suite of instruments.

Despite the present tonal weaknesses, I sense that electronic pianos will still become a routine purchase instead of a mechanical piano. Once they are improved and firmly established then the one instrument will offer not just a single piano, but a choice of alternative prime maker piano tones, a range of earlier fortepiano sounds, harpsichord, clavichord etc., organ and all the gimmicky imitations of other instruments and special effects. Some makers will already claim they do this, but in the same way Henry Ford said he offered a motor car for the masses, his reality was far inferior to that which is now of routine quality for us. My expectations of an electronic piano are set much higher than I have yet heard.

Electronics does of course offer a very wide range of other features for the keyboard. It is easy to transpose a piece of music into a different key, and one can combine electronically recorded music with the keyboard. For composition, there can be a direct electronic output that can be computer reformed as a musical score. In Chapter 6, I discussed the development, and differences between various musical scales which currently cause problems when a fixed keyboard tuning is played in parallel with the more flexible tuning from a voice or a violin etc. State-of-the-art electronic pianos are already potentially able to reduce these problems by adjusting the type of musical scale that is to be played (e.g. not just the familiar Western scales, but also those from more distant countries). Equally it is now possible to set a tuning to an individual taste. There are some major distortions of the piano tuning that are a compromise, and never totally satisfactory on a mechanical piano. In principle, these could be addressed in a future electronic version. The only caveat is that the resulting instrument need not sound like a piano. This is not a criticism, and I view such changes in the same way that we may choose to play some music on a fortepiano and other works on a pianoforte. New instruments are purely electronic and unrestricted by mechanical actions. Therefore, I envisage immense opportunities for new features. For example, the volume need not fade away, but could be controlled to swell or stay constant for a longer period of time changed, for example by a pedal control. Similarly, I have seen electronic keyboards offering intensity tremolo (as on electronic organs) or frequency vibrato (as with a voice or string instrument). This means the piano action will have moved up to the flexibility in volume that we have with our voice, or with string or wind instruments.

As to blue sky imagination of other features on an electronic piano, I am sure that many companies have teams of engineers considering such issues, and merely from writing about it I can easily imagine several futuristic options. Implementing them might be difficult, but in 20 years they may materialize. Their appearance will not be set by musical factors, but by patenting and marketing, as these are more important for an electronic piano company than for users such as us.

Electronic Instruments

In this book I am making several references to electronic pianos because they try to emulate the classical mechanical instruments, they are improving, and in principle a single machine could switch sounds between many different types of historic keyboard instrument. I see their future role as extremely important, but they are not the only electronic instruments to have been created. One interesting example came as early as the device patented in 1928 by Leon Theremin. It contains some metal plates, and frequency-controlled oscillators, which are changed by moving one's hands between the plate antennae. It needs skilled hand control, but offers an eerie (i.e. unusual) sound, quite unlike any instrument which we normally use in music. We may rarely have seen one in a concert hall, but the sounds are actually very familiar as they have been used in numerous films, and even a TV series (Midsomer Murders). The theremin is equally a good example of technology driving musical composition, as not only has there been the eerie mood music for films but composers as diverse as Miklos Rozsa, Dmitri Shostakovich, Bohuslav Martinu, and Percy Grainger, have written concert works for it. It also features in many of the more popular music group sounds in a variant called a Tannerin.

More familiar will be the wide variety of electronic synthesizers. These offer flexibility on sound production that is not feasible from conventional instruments. Their output is built into many recorded sound tracks. While we normally do not associate them with earlier forms of classical music the definition of 'classical' is quite dynamic and certainly the sounds are now familiar (or even essential) from many types of CD or radio programme. These, as for any type of electronic audio device, will benefit from ongoing advances in technology, and so they will become both more flexible and easier to use for composition. Hence it is inevitable that synthesized sound (and music) will increase in familiarity and be aimed at wider audiences. Probably the key step will be to have performers who capture the public imagination.

Other Technological Influences on Music

On the scale of technological advances over the last 200 years the development of musical instruments is a rather minor item. For most of the world, new music technology will mean the development of recording formats from records to CDs, films and television plus DVD variants, and of course broadcasting. Music is firmly entrenched in every aspect of these communication systems. It is used as background music to set moods in films or TV, to influence our buying habits in supermarkets and shops, and to make adverts have a jingle that means we remember them better. Music is a key tool in the psychological warfare of conditioning us both for entertainment and marketing. Realistically, there is little difference between a Wagnerian leitmotif to signal that a key character is about to say something in an opera, and a commercial theme tune to say we are about to watch a familiar TV series. The broadcasts and recording techniques have brought an immensely wide view of music from many cultures within easy reach of all of us. The systems also offer sounds from the most famous musicians in the world, and so this sets a reference standard of what we should expect to hear and how music is interpreted by our current idols. This is not always positive since live concerts or our own attempts at performance at home may not match the carefully manicured CD version, and this can be inhibiting and/or discouraging. Excessive access to TV, radio, and mobile phones may equally curtail our attempts to learn an instrument and enjoy music from our own efforts.

   The various media need music, and so it is a good career opportunity for many composers. However, composing for different media undoubtedly has a deeper and more profound effect on compositional styles for all composers of our era. Radio, TV, and cinema, with their associated music, only commenced significantly within living memory of many people. This makes it difficult to find an impartial and objective view of how this is shaping modern musical appreciation and composition.

Based on the speed of change so far, it is even more uncertain as to where we are being led by these electronic media and recording techniques. The only certainty is that musical styles, performance, and instruments will continue to evolve, and technological innovations will play a very major role.

CHAPTER 9

DECAY OF INFORMATION
AND DATA LOSS

Technological Driven Changes in Music
and Composition

In the preceding chapters I have emphasized that advances in the technology of instrument design have had a major influence on the styles of composition and performance. Since they were accepted and have persisted, the obvious assumption is that we like them, and see such evolution as being positive and desirable. On the other hand, it was fairly clear that the introduction of the equal temperament tuning was a compromise. It was forced onto the musical community because of the central role of keyboard instruments. This is because they cannot cope with the flexible tuning that is intuitively used when we sing or play on stringed instruments. The piano tuning is not adjustable for different keys, or modified when using different chords, or with variations in ascending and descending sets of notes.

We also saw some very valuable additions to the survival of music once we were able to find schemes of notation that could produce music manuscripts and remove the need to hear and memorize works from other performers.

What is less immediately apparent from a focus on these positive features is that technology, by offering change, is simultaneously a force that means we lose track of earlier musical styles, presentation, and understanding. This is not just a problem for music but is equally obvious in all systems that try to offer a record of historical writing, painting, or culture. It is easy to delude ourselves that because we have documentation, we have understood the past. It is also apparent that, just because we have pictures and carvings of ancient instruments, we still have no idea how they sounded or were played. If one doubts this, then consider looking at images of modern foreign instruments or manuscripts that are not used in our local musical world. We will make very poor guesses at the sound, and totally miss out the underlying patterns and flexibility of interpretation or be ignorant of the text.

The one message which I am trying to offer is that technological progress has many benefits but it is a double-edged sword, and the very act of progress, destroys our understanding of the past. To justify this unfamiliar view, I will briefly take some examples from language, writing, and painting. In a later section I will then address the way recording techniques have supported how we make totally different interpretations of music, even within the timescale of such technological progress.

Survival of Literature and Painting

Music is frequently discussed and compared in terms of art and literature of the same period, so we should also consider how well they have survived over the years. Survival is not just in physical terms, but also in understanding of the cultural environment of 'then and now'. Visual arts have an immensely long-term history. There are cave paintings dating back many thousands of years, paintings from ancient Egypt, sculptures from Greece and Rome etc. as well as equivalent items from many other parts of the world. The paints used were initially very simple, with a limited range of colour, so technology has helped to develop better versions. Nevertheless, technology has not given any insight into why cave paintings were made. Historians and sociologists make guesses, that they had religious significance, that in some way either deified the animals, or made them easier to hunt. Many views are feasible, but quite disparate, or maybe none are correct.

Art shares, with music, the feature that works are original and produced without a formal grammar or spelling, so everything from the composition of a work to the way it is seen (or heard) is highly subjective and requires us to add in all the effort needed to interpret it. We may benefit from explanations and views of others but in the final assessment it is a judgement based on our experience and sensitivity to what is on offer. There is no absolute standard.

Paintings and literature both suffer decay of the media on which they are drawn or written but, at least for literature, the more popular works may have existed in many copies. Artworks that have been copied will not be identical to the original. Changes in composition and quality are inevitable, not least as the skills, concepts, and materials of the copyist may differ from the original. A classic example of alterations in design and quality is offered by the Leonardo painting of the Last Supper. It has been critically discussed by art historians, although only copies have survived. Realistically their perception is influenced by a mixture of distorted and restored originals, and their imagination based on copies. Technology is an unfortunate factor in this case, because of the choice of paints and background material, which did not survive well. The Last Supper original was painted by Leonardo da Vinci between about 1494 to 1498 on the north wall of a damp building, next to kitchens, and in a room later used for many purposes, including billeting troops. By about 1642 the original had virtually vanished. Restorers have changed the colours, redirected the gaze of some people, and altered the background. This is the musical equivalent of a waltz for soprano, transforming into a march for a bass, but with the same title.

Other paintings of that period, such as The School of Athens by Raphael in 1509, are still in excellent physical condition. Nevertheless, a modern viewer will totally miss all the symbolism of whom the figures are supposed to represent. Also, the mythological and allegorical implications of the statues and gestures by the characters in the painting may convey nothing to a modern audience. Art historians are not in total agreement, even though they have familiarity with the period. Information exists but it is encoded in ways we cannot interpret.

Discussion of the survival of literature will of course highlight how changing technologies have influenced what can be written, copied, and distributed. The focus is usually on the improvements and rarely on the weaknesses of such change. A quick review of literature examples would have to range from runes carved on stones and clay tablets to parchment or vellum, up to varieties of paper and printing. A modern perspective will include electronics via computers, and all the associated electronic access and storage alternatives. For the more rugged materials, their physical life expectancy is several thousand years, and even parchments, such as the Dead Sea Scrolls, are sometimes readable (just) after some 2000 years. While the artefacts have survived, many of the languages have vanished or transmuted into some later equivalent. Surprisingly, loss of the language has rarely been a total barrier to translation.

Pictograms on clay tablets from 5000 years ago in the Middle East evolved into the script called cuneiform. This code was broken by a German school teacher (Georg Grotefend) who found an inscription relating to the kings of Persia, and a trilingual inscription written in languages of early Persian, Elamite, and Akkadian. A better publicized example from an Egyptian site was the Rosetta stone. The Egyptian hieroglyphics were written for Ptolemy the fifth in about 196 BC. They were carved on a basalt stone with the same information in Greek, Demotic, and Egyptian hieroglyphs. This single piece of stone was a key to decoding such symbols. Even without the multilingual help of ancient government decrees etc. other scripts, such as Linear A and Linear B, have been understood, and just needed a persistent and inspired translator. Linear A was a mixture of syllabic and ideographic writing found on Minoan Bronze Age pottery in Crete, and dated from around the 15th century BC. The Linear B was on clay tablets and was written somewhat later, and is now assumed to be the early version of Mycenaean Greek. All these examples show that written material was available from at least 4000 years ago. In the Far East there are Chinese inscriptions on jade and clay dating from nearly 3000 BC.

In the case of understanding very ancient music and how it might have been performed in various regions of the world, we have no equivalent Rosetta stone to contrast music of the same period in different countries. This is unfortunate, but a positive view is that we have a medium which is not limited by language. Therefore for music we have avoided one of the difficulties of conventional archaeologists.

The problem of information loss is not confined to a retention of the original writing and translation but is just as dependent on knowing the cultural context of the time, or how the meaning of the words and symbolism were used. For a modern English speaker listening to a play or reading a script by Chaucer, Shakespeare, or Dickens we will understand most of the words, probably miss the fact that many words have shifted in meaning and nuance, and we will definitely fail to understand topical references, jokes, and the social backgrounds of the writers and audience. We may grasp the thread of the plot and enjoy it from our viewpoint but this is not the same understanding as was at the time of the writing.

I have mentioned how background knowledge is equally important in music, as typified by the operettas of Gilbert and Sullivan. For us, we seriously miss out on the music and text of that time. Much of the music was a distorted lampoon of other Victorian music and performance styles. The texts were even less subtle, as they were very pointed comments on life and political acts of the late 19th century. Even though we know this, there is no way the operettas can offer the same impact as they would have done to the Victorians.

This steady decay of not just the materials but also our understanding of the culture and opinions of the original audiences means that we have the same ongoing problem with music. We may have the score and documentation about the period and possibly the instruments that were in use but we can only interpret this in terms of our own experience, and we lack any acoustic examples of the original performance and the musical knowledge that was brought to it by the audience of that period. Adding the technologies of writing musical notations and mass printing of the scores does not reproduce the music of the time of writing. That is already permanently lost.

Film and Video Survival

Photographic film recording first appeared nearly 200 years ago. Photographic historical records show that there were steady improvements in techniques and sensitivity, from sepia to black and white, and then to colour, plus development and use in silent moving pictures as well as with sound. Modern cinema is just the current end point of this long chain of technological evolution. Films are important for music as they have recordings of performers, musicals, and operas that give a far better historical record than would ever have been possible from written descriptions. This is so, not least because many styles would have been the norm, and therefore would not have required or attracted comments from the reviewers and critics of the time. Survival problems are equally apparent because in writing, art, or sound recordings there is a similar impermanence of information records.

Early films used changes in the chemistry of silver halides when exposed to light, and the halide crystals were embedded in an emulsion. Unfortunately, neither material is totally stable, so all photographic images are changing with time. Colour images and prints seem particularly prone to change but this may only be partly true because in reality we are just more sensitive and less tolerant to colour changes than we are to fading of sepia or black and white images.

The leap forward from static photos to moving pictures at the start of the 20th century was exciting, and movement totally disguised the fact that individually, each frame of the film was of relatively low quality. Humans are really good at coping with poor signals so this did not upset us too much. (For example, we drive at night in heavy rain where we actually see just a few per cent or less of the view we would have in full daylight.) One major problem for survival of the early cine films is that they were recorded on a cellulose nitrate type film, which is not only chemically unstable and prone to crumble but can also react and become highly inflammable. This is a natural death route for the film. Additionally, the film studios did not keep old records and, with the advent of sound films, they often dumped all the older versions, not just because they were obsolete but also because the silver content of the film was valuable. Subsequent film compositions were more stable, but changes in film format and methods of adding sound tracks has caused obsolescence, and even in commercial films the colours of the master films fade or change very noticeably within about 10 years. Film records are therefore rather short lived. This is unfortunate, not just for dramatic or historical films but also for musicals and images of dance routines.

The BBC Doomsday Project of 1986

One classic example of signal loss for text, vision, and music is neatly demonstrated by the attempt by the BBC to make a modern equivalent of the Doomsday book. Back in 1086 the Normans made a very accurate record of all the people and property in the Britain that they had conquered. Their aim was of course to raise the maximum tax money, and legitimize their redistribution of wealth from the Saxons to the Normans. But in so doing they left a highly detailed record of life at that time. To celebrate this event, 900 years later the BBC commissioned the production of some interactive videos that could offer a snapshot of modern Britain. It was intended to be a valuable source of information for many years to come. In terms of the number of images and video clips, the challenge in 1986 was to find electronic writing systems, software, and hardware to record and access such a wealth of data. No standard media existed at that stage which could handle it, but the aim was to make it usable with BBC Microcomputers, which had been introduced into schools by a government grant. These 1980s state-of-the-art computers had no built-in storage disc and the memory was typically 256k (actually only 220k after formatting)! Therefore, an entirely new computer system was built for this Doomsday purpose, and the video material was stored on some extremely large discs. Unfortunately, the system was very expensive, and government funding for such equipment to schools had ceased by the time the project was completed. The copies did not sell well because the price in 1986, was roughly equivalent to that of a small car. Further, this entire process was clearly at the very forefront of 1986 technology, and it was not the direction taken by mainstream developments. By the mid-1990s there were thought to be no systems available to read it, and the computer formats were not compatible with the 1990s technology, plus the master magnetic storage tapes and discs had seriously decayed. Effectively the entire Doomsday project had been lost in a decade.

In 2002, the situation was partially redressed because some equipment and relevant technicians were still available. However, even for such a major project the Doomsday example indicates that technologically dependent information storage can be very transient.

All these examples underline that our enthusiasm for new technology and change with frequent replacements of the electronic systems means that electronic decay rates for information are probably no better than our own memory loss with time and age.

Technology and Recording

Science has come to our aid over the last 150 years, with a steady progress in recording techniques that approximate the sounds and interpretations of music by different performers. It has offered us access to music from across the world, with examples by the leading exponents of every type of music. As a scientist I see this as very positive but on reflection I sense that continuous easy access to top performers actually undermines our expectations of many other performances. It may even discourage some from being musical practitioners because their performance standards will never meet the idealized recordings. Further, excessive and continuous background from radio, internet, and mobile phones just become musical wallpaper. We are far less likely to focus our full attention on the music, which can then easily sink into a pleasant background experience. Once we hit this level of listening it is very hard to then readjust to fully concentrating on the music, rather than the other distractions that are going on around us.

The Recording of Musical Sounds

In principle the problems of only having printed manuscripts of music, and instead having the ability to hear the recorded sounds of earlier performers started to improve at the end of the 19th century. For the piano there was a technique of encoding performance on holes punched in rolls of paper, and these pianola (or player piano) recordings are reproduced on effectively the same instrument as that used for the production. In the playback mode, the holes controlled air pressure on a pneumatic system of pipes that operated the keys in a modified piano. Consequently, the instruments and their paper rolls offer some good insights into the phrasing, speeds, and interpretations of piano music played more than 100 years ago. By about 1900 there was an agreed industrial standard for the types of roll, and thousands of items were recorded by high grade artists for many types of music. The fashion and usage of player pianos lasted until the mid-1920s, when broadcasting and electronic recording appeared. Piano music from these instruments has been preserved, and there are also many collections with archives for such instruments that have allowed reproduction and transferral into more familiar modern media. Nevertheless, performers were not universally agreed that the recordings were faithful monitors of their musicality. It is still true that even with the most careful modern recordings performers say that they do not hear what they thought they sounded like (invariably they assumed they were better or more central to the music).

Although player piano design started in the late 19th century it is also very typical of how technologies have come and gone. For the pianola the development took around 20 years before there was a universal standard. Usage then continued for a further 20 years until it was effectively obsolete. Such patterns clearly emerge with later recording techniques and storage media. From the limited examples we have, the timescales of development, usage, and being superseded all seem to have been in similar length steps of say 10 to 25 years (i.e. the time it takes for a new generation to emerge). The new consumers will therefore buy state-of-the-art equipment, but if the items are expensive they may become trapped into continuing to use such items, rather than continuously updating.

Voice and instrumental music recordings were considerably inferior to the mechanical pianola technique, and although there were several 19th century variants of waxed cylinders, discs, and drums, they needed a very intense sound to impress recording into the wax. Some of the opera stars with more powerful voices made impressions, but realistically the recording arrangements were so different from a real performance that it is totally unfair to judge the singers by the quality of their recordings. The other problems of these early attempts were firstly that only one (or very few) recordings could be made at a time (i.e. no possibility of mass marketing); secondly, the act of playback seriously degraded the recording by wear and contamination with dust etc. Sales images of people (or a dog) unable to distinguish between original and recorded performances were certainly an extreme advertising exaggeration.

A rapid scan of progress in communication and recording might include dates and items as follows. My dates are approximate and they will vary between demonstrations and items that might have been marketed, but broadly they show how changes were evolving.

I will start with the efforts of a little-known inventor, Édouard Léon Scott de Martinville. In his patent of 1857 he used a horn to focus sound that vibrated a stiff bristle on a surface coated with lampblack. The vibrations scratched the surface, and the recording region was tracked via a hand-cranked cylinder. Recording was possible but playback was destructive. Nevertheless, in 2008, US scientists imaged and made a digital copy that was playable, and they heard him singing a fragment of Au Claire de la Lune. An 1860 example includes an Italian singer.

By 1876 there were demonstrations of telephone communication; this inevitably was a spur for considering recording systems, and by 1877 Thomas Edison managed to record himself saying 'Mary had a little lamb' by an impression made on a tinfoil cylinder. A wax surface recording then followed. The results were sufficiently successful that such devices were used by people to record native speech, and folk songs, not just in the USA, but in many other regions from Europe to Russia. By 1891, Emile Berliner made a big step from the cylinder systems to recording on a flat disc.

Before becoming excited that all this was positive technological progress, with the good of music and preserving native culture as the top priorities, we need to take a more careful view of the reality. Edison was primarily an entrepreneurial and successful businessman. He was trying to sell his recording systems and used singing promotional tours as part of this activity. He claimed that one could not distinguish between the singer and the pre-recording. The trials were apparently staged in the dark, so that the audience could not see which was in operation. The sounds we now hear from his various systems definitely do not sound like singing, but to a naïve audience, who may not have been to concerts or heard professional singers, it was definitely an exciting experience. Further, he trained his 'singers' to sound like the recordings, when they did their part of the act in the light. In marketing terms, quite brilliant.

Similarly, many of the recordings of native singers, for example of poor US plantation workers, were only made if they sounded suitably primitive, and any singer who wanted to choose other songs, or sounded too educated or skilled were rejected (this did not fit in with the image they wanted to present).

Music Scores as We Have Known Them

If I return to the far more primitive technologies used for notation and printed music, then we see that initially these technological advances were fantastic, as they removed the need to memorize all the music that was to be played. This was not just for simple singing or individual instruments, but without it there could not have been any major instrument groupings or orchestras. For almost 500 years there have examples of printed music, and so it is hard to imagine the earlier situation. Printing quality and distribution has improved, often in consistent formats, so most popular works are available from shops, libraries, or the internet. This appears to be a stable position and, certainly for the publishers with copyright of the music, it is not a situation that they would wish to change. Nevertheless, no aspect of life is fixed and static, and for music the evolution of scores in new formats is inevitable. A general predictive view is that music will progressively become ever more available only in an electronic form, and this will offer many aspects and features which are both positive and desirable. By now, it should be obvious that I will also predict many changes that are undesirable.

Will Printed Scores Survive?

In the short term the answer is definitely yes, but it is far less clear in the long term how music will be available and be presented to the player.

I have shown that this is a general pattern for all types of information—whether in music, art, literature, science, or historical records—that information and data become lost or unreadable with time. In the case of music there is a further range of problems, so I will not only comment on them, but also offer some insights into why musical information loss is inevitable, even if data survive. In all cases, there are two key factors. The first is the survival of the hard copy of the music. The second is loss and change in our ability to understand and interpret what is being presented to us. Music has always had this double problem of changes in the tangible views of the written scores or notes and the emotive changes in the way we imagine they should be played and interpreted. An overriding factor is that it is impossible for us to accurately comprehend the cultural situation and attitudes at the time of composition. In reality, societies change quite rapidly, and so we can never be the same as the original audience or performer. Even today if we look at older TV films, comedy shows, or documentaries, they will often seem like something from another world, and yet we may have seen and appreciated the original. If this can happen in just a few decades then it is very clear that we have no hope of honestly pretending we can understand music of say 1800, with the mindset of that period. As I mentioned earlier, we should recognize that in 1800 the critics and commentators were saying that they could no longer understand the styles and music from 1600. We change, but the problems do not.

Written music helped the spread of music, and partially offset the interference of wars, plagues, politics, and natural disasters. Initially, writing was used for religious purposes, whereas less sophisticated folk music may have benefitted less because the general population were normally illiterate. In more recent times, language, dialect, and by inference folk music have been undermined by radio, TV, and travel. We are fortunate that many formally trained reputable composers spent time and effort (particularly at the turn of the 19th to 20th centuries) in collecting native tunes. Such items were then included in later compositions or published as original folk items. Works were also documented by others who were not composers, such as Cecil Sharp in the UK. Luckily such efforts often preceded radio broadcasting, which homogenized our language, accents, and musical expectations.

The early versions of musical notation were very different from those in use today and, except for expert scholars in this field, understanding the written music of the past centuries is not possible. By the time of the baroque period the written scores were a minefield of potential errors and misinterpretation for a modern reader. Many of these difficulties were highlighted in a highly detailed and readable book by Robert Donnington, *The Interpretation of Early Music.*

Reading difficulties are not confined to ancient Baroque music, as errors can occur because of differences in timing notations and speeds by composers such as Handel or Beethoven. These have subtle variations from those in current use. The very modernistic notations or writing for electronic music are equally obscure for a standard classical musician. A review that spans both the ancient and future notation problems is included in the New Oxford Companion to Music in an article by Anthony Pryer.

To emphasize the scale of the differences I will give just a few examples. Before we begin to play we expect to read a key signature that implies which set of notes are relevant for the piece. Baroque signatures were not consistent with those in current use, and in some cases several flats might be indicated, implying one key. However, the work could actually be written in a different key(!) and this mismatch was adjusted by introducing accidentals within the score. Accidentals (i.e. notes that are not normally part of the scale being used) are a particularly dangerous area, even for the Baroque experts because, instead of our current symbols for flats, sharps, and naturals, they indicated changes to the notes with at least a dozen other symbols. The symbols did not necessarily mean the same thing in all situations but had a wide range of meanings depending on the piece being played. For example, there was not even consistency in writing the accidental *before* the note that was to be changed. If there were repeated notes, the accidental might appear in the *middle* of them. Bar lines were not always used, or they might appear in some parts but not in others.

Keys include both major and minor variants. For the key of C, the minor would include C, E flat, G, whereas the major would be C, E, G. The third is thus the obvious indicator of a difference between minor and major. In some Baroque music, if the work were in a minor key and it finished with a chord that included the third, then the last chord would instead be played as a major chord. This was standard practice, but not written in the music. Equally confusing is that it was termed a Picardy third, but was not in fashion in France.

Stylistic patterns and rules of composition were familiar and ingrained in the training of the professional musicians; so many composers did not trouble to write parts that they assumed would be automatically added by the performer. Classic examples are in say the bass part played by a continuo or organ, where the performer was expected to add a complete bass line developed from the occasional chord indicated by the composer. For top-grade performers such as Bach or Handel they did not need any written music to sense what was needed, but for later performers who have neither trained to play this way nor perhaps been familiar with the style of the in-fill, then playing only the written score will undermine the work. Equally, inexperience and an incorrect set of additions could also ruin the performance.

Embellishment and Improvisation

The critics and musical essayists of the Baroque period were no more united in their opinions as to what was good practice than critics and audiences are today. Nevertheless, it appears that many intervals and chords that we find perfectly acceptable were heard as undesirable and dissonant. We need to remember that they were using musical scales that were 'natural', or intuitive ones based on unaccompanied singing, and not artificially constrained scales set by the equal temperament tuning of the keyboard instruments.

Perhaps more obviously to our ears, there was a major difference in the way the music was embellished by the performers. This was not only for the continuo type parts, where the composer left the performer some freedom (but with the expectation of the 'standard' style of the improvisation). Other decorations included trills, turns (a note followed by up and down adjacent notes), acciaccatura (short notes crushed in before the main note). More confusingly, appoggiatura, notes slipped into the music, but with different treatment depending on the composer! Also, many other types of unwritten embellishment were introduced by soloists to show off their skill. Performers deviated greatly from or minimized the written score in favour of such personal input, plus they freely added showpiece cadenzas and other improvizations.

There is an anecdote attributed to Rossini, who heard one of his arias in The Barber of Seville heavily embellished and used purely as a virtuosic showpiece. He commented to the lady who had done this, 'Nice tune. Who wrote it?'

Top performers continued in this improvising vein for many years. It was expected by the audiences, and performances were often rated on the originality in terms of such improvizations. Even pianists as late as Liszt and Chopin were such showmen. Beethoven and Mozart were masters at original composition and improvisation. Less well known is that Mozart had a sister (Maria Anna Nannerl) who, while being a very able pianist, was not good at impromptu improvisation. To help her, Mozart composed musical improvizations for her (not in his normal style) which she learnt. In performance, she could then produce these 'spontaneous' additions to the standard scores.

In modern concerts the soloists not only do not improvise during the main sections of the music but they frequently use familiar cadenzas and only extremely rarely will a cadenza be a live improvisation. This is not to say that alternative cadenzas may not be acceptable; indeed I have heard some that were equals of the more familiar ones. However, I have also heard attempts to add technically brilliant cadenzas, in works such as a Mozart violin concerto, which were totally out of character with the concerto (e.g. advanced violin tricks of left hand pizzicato and stopped harmonics). For me the use of new cadenzas is fine, but they must musically fit the character of the main work. A truly competent soloist can have the courage to play simple music, simply as intended, and use pure showpiece items to demonstrate virtuosity.

An Overview on Loss of Musical Information

It is inevitable that music, people, culture, musical styles, and expectations of performance, improvisation, tonality, etc. are all transient. We are evolving because of exposure to an ever-widening choice of music, performers, and availability to hear them. This means that we are the experts of today, and arbiters of our own musical taste. Assessments are highly personal. Technology has brought us this freedom in ways which are mostly positive. Nevertheless, I personally sense that these same advances in access mean that we do not sit back and consider sufficiently carefully what we are hearing. Over-exposure without critical input, dulls our senses to our immediate enjoyment, and undermines our long-term musical appreciation. For passengers on a train reading, emailing, and simultaneously listening to music via very low-grade sound systems on tiny headphones is totally undermining their musicality. Technology in musical terms can be fantastic, but it is essential that we use it wisely; if not, we are the losers.

.

CHAPTER 10

FROM LIVE MUSIC TO
ELECTRONIC OFFERINGS

A Century of Electronics and Music

During the last century the opportunity to listen to music moved from the confines of concert halls and home performances to a wide variety of recorded and broadcast formats. This means that we can now enjoy and compare works from across the world, performed by artists both live and dead. The continuous advances in recording techniques are spectacular and span devices from the 19th century wax drums to modern CD, DVD, downloads, and streaming. Each step forward has been hailed as a major advance, starting from the 19th century advertising hype making totally untrue claims that 'one can no longer tell the difference between the original and the recording'. Clearly advertising in the past was not totally honest (still true) because on listening to those early attempts, and even the better records on vinyl disc or modern CDs, it is clear that there are still noticcable disparities between the concert hall sound and anything that can be delivered elsewhere. This is immediately evident in a normal home environment. The current CD sound is enjoyable, and often musically satisfying, but it can never offer the identical experience to the live performance.

To those of us who frequently attend live events this is obvious, but we are likely to be a minority voice because for geographic and/or financial reasons large fractions of those who enjoy classical music will not readily have access to venues where star performers and leading orchestras are appearing. It is therefore particularly important that we are heard, or market pressures will dumb down the electronically marketed music, rather than try to raise the quality of the recorded and broadcast sound. For popular music, where electronics is already present in the staged events, the difference in sound between live and recorded music will be less obvious. Nevertheless, the atmosphere and excitement are still far greater at the live events, no matter which genre of music we subscribe to.

The bulk of the technological progress has involved electronics, initially in the design of primitive microphones and speakers, to the amplifiers that benefitted from the invention of vacuum tube valves and to semiconductor components. Computer processing speeds and their internal storage capacity for data, music, and images are almost doubling every two years or so. These logarithmic advances with time continue to open incredible processing power. The performance of semiconductor circuits provides associated computing and storage possibilities, and are all ongoing, so they will continue to impact on the quality of broadcast and recorded music. The use of electronics and amplification etc. have entered into live concert halls for all types of music. It may be minimal for opera but it is often included in concert halls to offset problems with auditorium acoustics. This may not be obvious or universal for classical music but it is totally dominant in pop music. The effects range from amplification to synthesized sounds and real time autotune (i.e. electronic correction of mis-sung notes) and 'pitch bending' of pop music. The frequency distortions range from intentional changes for sound effects, before emergence from the loudspeakers in a live concert.

I want to summarize some of the ways that electronics has improved recording and broadcasting, look at features we may not be aware of, and also indicate that there are some trends in signal processing that are definitely not desirable. I have already indicated that there are danger signs in the signal processing of classical music, particularly because in terms of CD production, music sales, and audiences the classical end of the music is a minority activity compared with the much more extensive pop scene. The historic example, that in the 20th century the Beatles sold more vinyl discs than had ever been made for classical music, is a clear indicator of the relative markets. The electronic processing and distortions deliberately introduced into recordings and broadcasts are aimed at the much larger mass market but the processing techniques change classical music in ways that can be highly undesirable.

In part this is driven by economics by the recording studios, who, for purely commercial reasons, will focus on popular music. There is thus a potential split between different types of recording company. Similarly, many sound engineers are people who enjoy the pop music scene and they are unlikely to be aiming (or wanting or able) to generate the sound balance and dynamics associated with classical music. Further, much of the processing has moved into an automated phase, and the objective of the way sound is then handled may be inappropriate for many musical styles and differ from the tone balancing of the more classically biased engineers.

One unmissable difference between pop and classical is that pop is likely to be played at high volume in noisy environments, whereas classical will span a far greater intensity range. The pop recording and broadcasting aim for a high maximum volume, strong dynamic compression (i.e. the quiet parts are also made loud) and a variety of sound distortions to match the audience demand. Many of these processing steps are anathema to a classical music audience but the processing steps could be inflicted on them if the CD sound is effectively created via an automated process.

Technology and Music Processing Over the Last 100 Years

The start of the 20th century saw the first electronic devices such as microphones and amplifiers. These were the seeds that allowed recording. At the same time, they offered the power needed to drive loudspeakers. Quality of reproduction was not good, but immense effort from the industry has gently provided ever better systems. The introduction of so much post-performance signal processing has had a major effect on the final recorded or broadcast sounds. It also introduces a significant phase of sound control that may differ from the actual performances. A more positive aspect is that it can be used to correct errors made during the performance. Therefore, a very key part of the industry is the role of the sound engineers who monitor and adjust the signals that are broadcast or recorded. In live performances, as well as recordings, they are key participants. Their significance in the process is such that many CDs and discs include the names of the recording engineer.

This chapter will track the influence of the electronics and engineers and the changes that have evolved in their attempts to make analogue recordings where the patterns of the sound pressure waves are impressed into a recording medium. While this approach was the same as in the 19th century, the materials and quality of analogue methods have advanced greatly. The subsequent introduction and foibles of the digital recording will be discussed in the next chapter.

Microphones and Loudspeakers

As soon as microphones and electronics were added to the record-making process there were inherent distortions of the sound. They were less than those of early techniques, and therefore were acceptable. The microphones of the 1920s were simple and there were many developments to match different situations. Natural resonances occur over a limited frequency range, and in microphones there need to be support structures, so early microphones resonated at the frequencies set by the mountings. There are several designs in common usage. The details are not relevant here, but the type called a 'ribbon microphone' gives broad coverage, but the response is rather slow, so this alters the form of the starting transients (i.e. *precisely* the part of the sound that we need in order to identify specific instruments and voices). By contrast the 'condenser microphones' have a faster response, but their increased sensitivity means that if they are too close to a source with a strong transient, they over-react and give a harsh shrill sound. I have seen powerful sopranos back away from the microphone at high volume to avoid the ensuing distortions. Other variants (e.g. the dynamic microphones) have a thick diaphragm. Their weakness is that they over-respond to central frequencies, compared with those from low or high notes. Different types are preferred if the recording needs very specific directionality to pick up one instrument, or if it is needed for a wide angular coverage.

Since a recording session (or concert) demands both directional and overview sound, several microphone types will be employed. This poses problems for the sound engineers as they attempt to process and amalgamate the various signal channels from each source. We may not realize that our ears have a different sensitivity to sound from the sides in a horizontal plane compared with sound arriving in the vertical plane but unfortunately microphones do not mirror this directionality. So even at the very moment of recording with state-of-the-art equipment, we have always lost valuable sound information.

Microphones and loudspeakers effectively have similarities in design, and consequently related problems. For the loudspeaker, electrical energy is used to drive a cone structure to generate vibrations into the air.

The cone systems are highly inefficient in the conversion process, and perhaps only 1 per cent of the electrical energy emerges as useful sound energy. This means that the electronics must be powerful, which implies more energy, more heat, and inherently more frequency distortions.

Within the amplifiers, semiconductor devices have good efficiencies, often above 20 per cent. Nevertheless, the amplifier and speaker package will require considerable energy. In general, energy conversion between different forms is invariably an inefficient process. To place this in perspective, examples range from old style filament light bulbs which wasted at least 90 per cent of the electrical power as heat. Coal and oil sources rarely deliver more than 25 to 30 per cent of their energy as power or electricity. If a sequence of conversions is required, as in electric cars, each phase from battery electricity to drive the motor, to generating the original electricity source should be counted. In real terms this means that the car can have an overall efficiency below 10 per cent. This major drop in efficiency is marketed as progress!

The resonances of the cone structures in loudspeakers also mean they function over quite a limited frequency range. In a 'good' speaker system it is common to have a set of low, medium and high frequency components. All have different directional properties, efficiencies and frequency responses, and less obviously, unrelated phases of the sound waves.

If our music spreads uniformly over the audio spectrum, then we need amplifier characteristics that take the flat original signal and *distort* it to compensate for the different speaker performances. For a known and specific type speaker system this would not be too problematic because we can adjust the amplifier gain to offset the frequency-dependent failings of our known loudspeaker. Reality is that speakers will be very different in the homes of the various users. There are major problems in the frequency regions where the components overlap. Guaranteed compensation is then an impossible task for the sound engineer. The broadcast or CD is thus made for a 'typical' system (i.e. probably never ideal for any).

A minor feature arises from the phase of the waves that we are driving. Just like ripples on a pond, if we have two or more wave sources that are not in step then, on the pond, we can see interference patterns. From the loudspeakers we can sense similar sound wave phase effects, particularly for frequencies in the audio range where the wavelengths are comparable to the dimensions of our head. The signals to the left and right ear are offering us directional information which may or may not correctly relate to the original directionality of the signal sources. Phase effects are subtle, but for the enjoyment of music they are apparent, even if we do not consciously recognize why some systems, or positions in a room relative to the speakers, are better than others.

Rather than blame the sound engineer, we have to accept that, whether for broadcasting or making a CD, the corrections they try to impose on the electrical amplification signal to compensate for the final speaker system are an intelligent guess. Equally, the deliberate distortions they are adding will not suit all systems. Such difficulties can be quite obvious, and I have CDs that sound better when played on one of my systems rather than on another. The 'good' system depends on the CD (not the style of music or the specific piece). The variations in sound are not minor academic changes that one can only detect via electronic analysis but are real-world changes that we can clearly hear.

The Alternative Scenarios

No matter how we listen to music, we are faced with compromises, and it will become evident that we often have minimal control over the contributing factors. The alternative situations where we listen to music are roughly the following. The first is a direct and live performance with classical, or other types of musi, produced by orchestras, choirs, or soloists. The environments may be concert halls, opera theatres, or small auditoria, or even in our own home (palaces would be required for live orchestras!). Performance quality will vary depending on the performers, the acoustics of the rooms, and our location within these spaces. With experience and sufficient funds we can at least aim for our favourite seats for a particular style of music. A bonus is that, in all cases, we are hearing live and unique events. They may vary in quality but they will offer a real presence of participation in the music and the adrenalin of live action, not least as there may well be the occasional errors or differences between performances from the same group. Listening to live music is therefore always a compromise between a repeatable performance that is offered by a CD and an event where we and/or the players may be particularly responsive. Live is unique. We should not forget that when we are in a concert we probably listen far more attentively than we ever do at home, even when we are just sitting listening. In many situations the other distractions of home events will mean that we concentrate far less on the music, especially if it as long as a Mahler or Bruckner symphony. If the home or car sound is just background music then realistically the quality can be far less, and so will be our emotional response and involvement with it.

Sound Engineers, Live and Record Mastering

Sound engineers are an integral part of many types of performance, not just when there is a recording or broadcast being made. In live pop music where there is electronic amplification and broadcast via speakers, either in a room, hall, or outdoor festival, the engineers control the dynamics and balance of the various performers. The outcome may well be different from that of the group playing where they can directly all hear one another and no changes are externally imposed. Feedback to them by headphones will never match the sound they believe they are making.

In operettas, musicals, and many stage shows the singers and performers have small portable head microphones so they can move around the stage, and the output is again controlled, balanced, and modified to boost or reduce different parts of the frequency range. For a shrill singer the upper notes can be reduced; for one with thin tone, boost can be added, etc. The sound engineer is thus an integral and important part of the overall performance.

The electronics engineer may choose to mix and fade together the various components of the on-stage sound (and orchestra). For example, in a duet the engineer will often need to alter the tonal and intensity balance between the singers, and the singers may be unaware how this has been changed from what they thought was an ideal duet on the stage. Equally, the audience may not appreciate how the performance is being controlled by the one strategically positioned engineer. In general, there is also the inevitable conflict that the engineering skills, from a person who may deal with many types of music, will differ for each genre, and be coloured by their own musical experience.

The engineer must also ensure that the microphones are turned off once the performers leave the stage!

By contrast, in opera performances personal microphones are rarely used, but there may still may be sound amplification in many venues, and the balance between soloists, chorus, and orchestra will be altered during a performance. The sound engineer thus has an extremely important role in the music and not only requires expertise in the manipulation of the electronics, but also a very high level of competence in understanding and introducing modifications into the music. As will be obvious, this is a far more critical task than for a subsequent recording engineer, as in a recording the initial tracks can be retained and reassessed in collaboration with the performers and conductor.

For pop music, and of course any concert that is produced in the open air, the role of the sound engineers is particularly important, and there may be several people coping with the signals from different instruments, additions of electronic synthesizers, and also balancing the sounds that are broadcast to the open-air audience. The hazards of having a successful sound event means that in many cases the superstar soloist may actually be singing to a microphone that is not active, but instead a prepared tape is providing the lyric. This may seem an odd approach, but at least it guarantees that the sound balance has been correctly adjusted for the venue, and indeed the soloist will have had input into how it is going to sound. Problems of lip synchronization in open air events cannot be by direct sound, but will be via the video screen views, and the broadcast music from the loudspeaker system. Many open venues cover more than one hundred yards, so direct viewing and listening are irrelevant, as the time lag could be as much as half a second.

It may seem slightly disillusioning to realize that the performance has the star appearing from a tape, rather than live, but such effects have been common practice for half a century. In many of the earlier film musicals from Hollywood the stars who were playing the key parts, were chosen for their acting skills and glamour, but had singing voices that did not match the musical requirements. The studios therefore employed 'ghost singers'. In many cases the stars did not even know this had been done, as the singers were forbidden to mention it. We now know that we have often listened to Marni Nixon, Bill Lee, India Adams, and many others, from the 100 or so ghost singers. Undoubtedly the same still happens, but at least they may now have credits and royalties from the films and DVDs.

Directors and Camera Crews

The final two scenarios are ones where we no longer directly see the performers. These can be via broadcasts of radio or TV, or by recording systems such as a CD, DVD, or film. Broadcasts of live events will have all the same immediate and live input from the sound engineers and camera crew directors to focus our aural and visual attention on features that *they* think are particularly important or interesting. These are unique and individual decisions that can have significant influence both on the sound and on our concentration, from changes in image. Inherent to all the sound features will be the electronics of recording, amplification, altering tonal qualities, and attempting to modify (i.e. distort) the music so that, when it emerges in our home our local electronics and loudspeakers, it will give a fair impression of the event. One can be very critical of these intermediaries and their influence but in reality the greatest changes in sound may still come from our own home features and the room where we are listening.

Changes introduced by the sound engineers are hard to quantify. By contrast, visually we can develop a fair feel of how the performance has been influenced by considering the snapshots of views offered by the images on the TV. Inevitably they focus on the conductor, soloists, or people that appeal to the cameramen and director. They switch quite randomly in many cases, and their enthusiasm to change to different pictures is often not in accord with what we would have done if actually present at the event. Many artists have had bitter arguments with producers as they see that such camera views and angles are changing (e.g. the red 'on' light changes between cameras). Artistically, the switch of concentration between images can undermine both the musical and emotional impact.

For us, the viewers, these changes are outside our control and are distracting in terms of the sound. In this sense, radio broadcasts are slightly better from a musical viewpoint. Similar distractions appear for opera, and I have seen the same opera both live, and the next day on a TV broadcast. The narrow field of view and/or zooming for the TV version lost a lot of the interest from other activity that was relevant and taking place on the stage. The sound was of course very different (and often inferior). Film versions are probably better, both because we never see the actual event, and also because the directors are able to consider several alternatives of how to present the action, plus important stage asides will be included in the camera view.

Separate sound distortions exist when listening to music via a TV. With the advances in TV design to make very flat screens, it is impossible to include high grade loudspeaker systems. The makers admit this and recommend an addition of sound bars or headphones. They can cost as much as the TV, so this is a good marketing strategy, but musically annoying.

Can Recordings Give a Faithful Rendition?

By now it will be abundantly clear that the answer is no. The recorded and broadcast variants of a live event must be treated as different but related performances. Comparisons between them are unfair on the performers. Almost without exception a CD that includes soloists will have a dynamic balance that makes the soloist much more pronounced than one can ever hear in a normal concert seat. Flaws, coughs, unbalanced performances from different sections of the group or orchestra will have been amended, either in terms of intensity or tonal balance, and the final edited version will contain many such corrections. There are some artists who are remarkably free from performance faults in their playing or singing, but even for them the rest of the recording may need repairs for an idealized CD. I have read that it is not exceptional to have as many as 150 'adjustments' made on a classical CD. These can include patches made from different recording takes, as well as other balancing techniques. The apocryphal comment from the sound engineer to a soloist after many such repairs and retakes is 'Don't you wish you could play like this?'

The 'repairs' I am mentioning here are valuable to correct human error and problems of the recording hall or studio. But some changes in the sound of the music are the result of the recording process. Wax discs were clearly a novelty, but mechanically they greatly distorted the sound. Vinyl discs eventually had a much better frequency response but the mechanical motion of a needle cartridge on a long tone arm had natural resonances and had difficulties accurately reproducing the entire frequency range with identical efficiency. The hard needle also caused wear on the disc. Overall this meant that there was a background hiss that was obvious in quiet passages, and a sound distortion that we associated with vinyl recordings. By contrast it is feasible to have digital signals on electronic systems that have negligible hiss and noise and no inherent resonances or signal distortions. People are remarkably perverse and, rather than appreciate such purity and clarity of sound, many will say they want the vinyl type distortions that they grew up with and are accustomed to. This is not just an older generation problem, as I will mention that an equivalent desire for distortion is prevalent in the young generations (particularly of pop music) and the sounds are being progressively *more* distorted to meet this demand. Contrary to any intuition, there seems to be an increase in internet sales of vinyl discs. Even more bizarre is that it appears that 40 per cent of the purchasers do not have record players! There has also been an upsurge in tape recordings (e.g. the small market rose from 2016 to 2017 by ~35 per cent).

Hidden Factors in Record Making and Mastering

To try to hint at the way modern recording is made, and processed and mastered onto a CD, I will try to sketch the requisite steps. Rather than just two or three microphones as was done for early vinyl records it is now more likely that far more microphone channels will be used. In part, this is possible because the modern electronics and computing power is sufficient to cope with a larger number of inputs. For pop music, microphones may be used close to the performers but in a studio this effectively means that, for a small band, each instrument (or small group) can be isolated, as can the accompanying singer. The danger is that group performers who are isolated except by headphones say they can lack the close-knit performance contact that they enjoy in live music, and so underperform.

There are consequences of close microphones, in that the frequency balance may be shifted and there may be unexpected features such as that a singer or instrumentalist who is facing a metal music stand may have reflections from the flat surface. To a microphone, the two sources, instrument and reflection, will provide waves that are not in step and they will beat together (i.e. like intersecting waves on a pond) and overall this can produce a fluttering sound that is transmitted as a real signal. The solution is simple: an open wire music stand or to play from memory.

The various microphone channels are then recorded on a multi-track tape system so that they can be processed independently. Editing here will normally be done after the signals have been amplified so that there is a substantial signal to work with. The objectives are, as always, to remove mistakes and extraneous noises, either by patching in an alternative version or by substituting a totally newly recorded section. For pop music, there is frequently the pitch bending option that I mentioned earlier.

Once the correct package of channels has been assembled, further treatment is needed. The sound signal from the microphone comes in as an analogue signal (i.e. the wave is recognizable as the pressure wave that hit the microphone), but this is now converted to a digital signal by sampling the wave at regular intervals to see if there is any signal or not. The sampling rate needs to be faster than the highest frequency that we want to use. We can hear (when young) from say 20 Hz up to 20 kHz (20,000), so we need a sampling rate of at least 40 kHz. In fact, a typical unit may operate at 44.1 kHz, which is fine for the top notes in many systems. Nevertheless, higher sampling frequencies are preferable because when the engineer tries to reduce noise or combine different channels, there are fewer false electrical signals generated. So, higher sampling and processing rates of say 96 kHz may be used. At this stage the sound engineer ceases to be a mechanical minor role player and instead becomes a key person in the record making. The tone balance between instruments can be adjusted by frequency filtering and amplification, the background noise level can be suppressed, and the balance between instruments adjusted. For pop music it is not uncommon to finish this phase with two or four compressed tracks from the backing players and then to totally re-record the lead singer to add to this adjusted background. This is the stage where music from synthesizers is usually added to the sound tracks. In engineering terms these steps are not always totally compatible, as the digital rates from different sources may not be the same, so skill is needed to overcome this.

With all types of music there are further frequency dependent distortions added in the amplification needed to drive the loudspeaker systems and offset their variability with frequency. A typical loudspeaker will change efficiency by ~12 dB per octave (i.e. the loudness will roughly halve, or double, over one octave) and this is definitely unacceptable, so a correction (called equalization) is applied.

Several techniques are used in pop music which (at least to me) seem inappropriate for classical music. The first is that if the CD is a compilation of different songs and items then the practice is to adjust their volume levels so that they are all of a comparable loudness. Our hearing range is extremely good over a very large span in loudness, so this seems totally unnecessary. The classical equivalent is to suggest that, in an opera, all arias, choruses, and recitative should be in the same narrow intensity range. The same is done to the tone quality of each item. Once again this is undesirable. Even if the singer is the same, but the backing was from different groups at different periods of time, then I would want to hear these variations, to contrast how the music has changed with time, or by association with other people.

Because of the perversity that people like the distortions of vinyl recording, many CDs are over-amplified so that the signal level saturates. This alters the harmonic content of the notes and boosts the higher frequency components. The two main effects are that the tone has changed and that notes all sound alike. I have heard many re-monitored compilation CDs (both of classical and popular music) where every track has been adjusted to the same dynamic. These interventions destroy our ability to follow changing development of the artists, as the sounds have been deliberately suppressed or homogenized.

Finally, there is a general trend that the dynamic intensity range is heavily compressed. Loud notes are driven up to the saturation limit, and weaker notes are amplified even more so the total dynamic range is limited between loud and very loud. This may be the desire of those listening to music as background in noisy environments, but musically it is rubbish. It has lost all sense of subtlety, and all items seem similar. For classical music this approach should be strongly resisted.

To place the concept of intensity compression in perspective I will repeat that at best we have a natural sensitivity range of over 100 to 120 dB (i.e. more than a million to one in terms of intensity); FM and vinyl can manage perhaps 60 dB, but a CD is potentially excellent at say 90 dB. This is sufficient because background noise in a quiet house will be ~10 dB and so the CD must be louder than that. However, the advances of compression cut all this musical expressiveness back to around 30 dB, which is no better than a radio in the 1930s.

I am expressing considerable concern because classical music is not the greatest market for CDs and many of these processing and disc mastering steps are now so familiar that they are becoming automated. Once this happens, the same dynamic compression and continuous high volume will be forced on the classical CDs, together with the deliberate distortions that come with high intensity saturation. Once the music is recorded in this way one cannot undo the problem by playing at a low volume setting: it will just be distorted low volume, and this has already destroyed much of the subtlety and musicianship of the performers. This is true for all types of musical event.

In general, pop music is compressed to a very small intensity range, so it loses subtlety in its dynamics compared with the actual performance. The only good factor here is that much of the rock/pop style music is delivered by microphones and electronics. This means that the 'live' performance sound can be much closer to the recorded version.

Future Electronic Screen Displays

One further electronic device that is of potential value to performers is an electronic screen display of the music. Quite clear advantages are feasible: for example reading music on a computer screen could easily be linked to a feedback sound detector that recognizes which section is being played, and we would therefore see a steadily rolling dynamic display of the music. It would be simple to control how far ahead was on view, and it would totally eliminate the ever-present problems of page turning. Another desirable features would be to allow control of the print size. So, depending on lighting conditions, eyesight, and complexity of the music, we could pre-programme it to offer greater clarity whenever it is needed. Adding notation on details of expression, fingering, or bowing etc., or pronunciation of foreign text, would similarly be both possible, and selectable, as we update our thoughts when we experiment with the optimum choice to suit our personal skill. Inclusion of metronome style indications (e.g. a pulsating visual marker) would be a trivial addition if we wanted it. For orchestral or musical groups, we could highlight critical features, such as entry points for an accompanist, or ensemble members.

The benefits of such an annotated, and automatic, page turning for a conductor are particularly obvious because conducting from a score requires frequent page turns, and the print with many orchestral parts is rather small. An electronic display could be selected to highlight any sections that the conductor needs to concentrate on.

In terms of economics, a computer display for an orchestra could be justified because a purchase of the multipart scores would offset the high rental costs of musical parts. Electronic copies of an orchestral work could easily be updated by a conductor, not just to highlight entry points and dynamics, but also to have updateable annotations that say what is needed from different sections of the orchestra. Since orchestral members and conductors are now drawn from many nations, any written comments on the electronic score could be in the language appropriate to the performer. This would benefit those struggling with language difficulties. An example of a Japanese conductor addressing Chinese, German, and Italian players in English is not unusual—often partially successful but sometimes far from perfect. I recognize precisely the same difficulties from international physics conferences and research laboratories. On one occasion, I was introduced as the only speaker not using the language of the meeting, which the chairman said was 'broken English'.

Some of this technology is available, but not at the level that I envisage for the future. The advances proposed here, for electronic display options of musical scores, are well within the scope of modern electronic and computer technology. There certainly would be a market, and it seems to be an inevitable opportunity for a new product with worldwide appeal. There are further benefits. Singers rarely fit into a neat classification of their optimum voice range. For best results they may prefer to make transpositions into different keys to suit the range of their voice. Skilled professional accompanists can do this by reading a score in one key, and switching the performance into another. Most amateur players find this very hard, and the act of transposition undermines their concentration on the actual musicality of the performance. A direct score transposition option for the electronic display of the music (rather than the notes being played) would open a market for home use. Importantly, this would work for all instruments. Current transposition is possible on electronic keyboards, where we read and play the original score, but the instrument produces different notes from the onea that are fingered. It works perfectly well for many, but it can be very disconcerting for players with absolute pitch, who realize that the note they are reading, differs from the note that is being played. Some of these key change features are already available, to a limited extent, on recent music writing software, and they could be made compatible with the new music display. The only weakness is that one would be reliant on computer software and their displays, instead of the highly reliable paper music scores. For a large orchestra it would require power to every music stand, so there will clearly be many situations where 'real' paper music will be retained.

These transposition options, speed control, and indicators of where we are on the score already exist on many internet download systems and are useful for learning new music, especially for singers. An unexpected downside of continuous texts and scores on a computer display is that they are not paginated in a fixed form. Scrolling, apparently makes it more difficult to memorize the work, as we lose a feel of where we are in a text and chapter. We use the visual information of the position of the notes and text to aid our memory (e.g. we recall that our key paragraph, or diagram, was in the middle of a left-hand page). This is not just a problem experienced with music, but in the recall of all writing and images.

Modern composers are already using software packages that allow them to write directly to electronic storage and display. With increasing interest in synthesizers, and electronically generated sounds, many of these scores will include features that can only be produced by electronics. The extreme version of this route is that some composers will decide they have no need of orchestral players, and their entire works will be electronic. It is not music as we now know it, but it exists in several music genres; for example, some musicals rely totally on keyboard generated support.

CHAPTER 11

ANALOGUE OR DIGITAL RECORDING

Vinyl

Sound reaches us as a time-dependent pressure wave, and to make a recording we need to capture this pattern. The 19th century approach was very direct, and the pressure wave moved a stylus in a soft wax surface and wrote a replica of the waveform. This analogue approach is simple to visualize. Early in the 20th century, with the invention of microphones and electronic amplifiers, the power delivered to the stylus was much greater and so the same technique was applied to writing a wax master that could be metallized and used as a stamp to mass-produce records. We had a record industry. It was not perfect because the final disc differed from the original sound because of frequency and intensity distortions inherent in the microphone, electronic amplifier, and mechanical movements of the writing machinery. Nevertheless, it was recognizable as music, and since people had never heard the original, they assumed it was an accurate copy of the performance.

By the middle of the 20th century the record material had improved to being vinyl, and they were longer-playing (LP) discs. These dominated the music market and their sales peaked around 1980 with 2 billion across the world. Discs were in competition with small compact magnetic tapes, and then suffered a major displacement by the advent of compact disc (CD) digital recordings. These discs also have inherent distortions via the recording and processing, but benefitted from steady improvements of the electronics and speaker technologies. The net result is that they were potentially a more faithful representation of the original performance.

Human nature is perverse, and many people claim they prefer the sound of the vinyl discs (when they are free of scratches and dust hiss), and that the original performance has been destroyed by chopping it up into blocks of sound that can be digitized. Preference is understandable. For the generation who grew up hearing vinyl, this is their musical language. In speech, we may all use English in the UK, but the pronunciation, phrasing, vocabulary, accent, dialect etc. will depend on our experiences when we were young. This is nostalgia of what we were accustomed to when young. So those of us who were raised listening to vinyl are highly likely to prefer it to any later invention. These are the types of sound distortion that are familiar.

The second comment, that the music has been destroyed by chopping it into segments (i.e. digitized) is false. By mid-20th century the recording for vinyl was captured on very high-speed magnetic tape (exactly as for making the later CD). This by definition is not an analogue signal since it relies on changing the direction of magnetization on the tape (i.e. it is effectively a digital system). Reluctance to accept the digital sound is interesting, as we are perfectly happy to watch TV, or view a computer screen, where the images are totally pixelated, and we rarely complain that pixilation has destroyed the images we are viewing.

Once we understand some of the technology of the recordings, and their problems, then we can appreciate why music sounds different on each format. Initially, the gramophone discs impressed a sideways moving pattern on the wall of the spiral recording track. Low notes cut in more deeply and limited the density of the spiral (i.e. short records). To compensate for this, the intensity of bass notes was suppressed, and playback amplifiers had more amplification for low notes. In principle, this redressed the recording loss. As usual, reality is different: it just meant the sound had changed. Once the recorded version had been heard a few times then we mentally adjusted to say that this is what we like and expect. The finer groove widths of lower rotational speed discs have other problems and again, in both recording and playback, frequency-dependent intensity distortions are actively used during recording. High frequencies may be deliberately suppressed, on the assumption that the speakers will not produce the sounds, or the audience will no longer hear them. Electronics and speakers, plus the rooms where we listen, all introduce further distortions, so it is hard to decide what is the ideal information to retain. Pop music is compressed to a very small intensity range, so loses subtlety in its dynamics compared with the actual performance. The only good factor here is that much of the rock/

pop style music is delivered by microphones and electronics. This means the 'live' performance sound can be much closer to the recorded version.

Gramophone discs changed in composition, weight, and storage capacity and in the earlier versions they were large and cumbersome, offering about four and a half minutes per side when running at 78 revolutions per minute (rpm). However, they improved over about 15 years and then were in fashion for a further 15 before they were displaced. Rejection of the 78 rpm discs came with the introduction of finer groove records running at 45 or 33⅓ rpm that appeared around 1948. Originally these could manage 22 minutes per side, but with different packing density of the spiral track, this was increased to above 30 minutes.

These flat disc records were not unchallenged, but were in competition with other recording formats, such as magnetic tapes. It is sometimes assumed that the frequency response of the vinyl discs is inferior to that of the later compact disc systems. In reality, for the better records this is not true. Their main weaknesses are the problems of wear, caused by the needle, and surface noise from dust (e.g. some hiss when the music is quiet, or clicks from dirt specks on the surface). They are also bulkier than the later CDs. The criticism of sound quality may be based on the fact that earlier records were often heard with reproduction systems that were inferior to those that are routinely available today. Good vinyl discs played with top grade amplifiers and speakers can still offer excellent sound. I repeat that some people prefer them.

The sales have depended on many factors; for example, shellac discs sold well up to 1929 but then plummeted during the depression and World War II. The invention of jukeboxes in the USA (late 1930s) created a boom market for one-third of record sales at that time. Vinyl was first invented around 1931, but was impracticable until around the late 1950s. The absolute peak in all types of disc sales came in ~1980, with 2 billion made worldwide that year. The one clear pattern is that there is an immense universal desire to listen to music of all types and cultural backgrounds, so no matter what format and marketing are in place, we will listen.

Magnetic Tape Storage

Magnetic tape storage does not fit neatly into the pattern of a technology that came, dominated, and disappeared. Certainly for home consumption recordings it made a major impact on the vinyl disc market as the units were small and portable but it suffered from stretching of the tapes and often there were mechanical failures for tapes that had been frequently played. The frequency response of the tape recordings varied with the speed of tape movement, and high speeds were needed for higher frequency fidelity, which of course was in conflict with a compact tape and a long playing time. Tone quality from the tape was not critical for the portable systems because the speakers were normally very modest in response, but in a better high-fidelity playback system the sound from the tapes was in direct competition with vinyl. Similarly, as with the vinyl discs, there was a background hiss in quiet passages. The sales importance of the home-play tapes was therefore from say 1950 to around 1990, when they were displaced by CD formats to become a very small ongoing market.

It is possible to follow the life cycles of the alternative systems by looking at their relative sales. By 1983 tape cassettes had exceeded the vinyl disc sales, but by 1994 (i.e. just a decade later) CDs had 80 per cent of the market, cassettes about 18 per cent and vinyl was down below 2 per cent.

Nevertheless, magnetic tape recording for the many microphone channels used in making both vinyl discs and the later CD recordings was and is a major part of the recording process. In the recording studio, the tapes can be operated at the high speeds needed for high frequency fidelity, and with wide tapes there are many operating channels (as many as 48 tracks have been used). This is ideal for the recording engineers and for editing of the recorded sounds. Magnetic tapes have therefore become the basis of the initial sound recording before editing into the marketed vinyl, tape, or CD.

For video, changes to electronic or magnetic storage tape repeated all the same problems of a media with a limited lifetime, as the magnetic tape material can stretch, and it does not have long-term stability, particularly if it is played frequently. A classic example of commercially induced obsolescence occurred for video storage on a system called Betamax which lost out to an alternative format called VHS. The reasons included the fact that the VHS could record a 2-hour film whereas the Betamax could not. In terms of equipment and image quality the differences were less obvious, with proponents for both systems. Each system survived for say 10 years before being superseded. A further problem with all such systems was that they were often used for recording television. Television formats are not static and are changing in terms of the number of lines, the dimensions of the screen, picture definition, and the way signals are encoded. These are very lively and active sources of change that are generally incompatible with earlier systems. Once again, life expectancy is probably not much more than 10–15 years. Equipment may be maintained to show home videos, but such recorded memories are unlikely to usefully reach the generation of grandchildren.

One should of course mention that for visual storage systems the electronic equivalent of the CD is the DVD. While the sizes of the discs are the same, the DVD packs some 26 times more information on the disc. In other respects, the physical survival problems are probably similar in terms of degradation of the disc material. This is not necessarily the only factor to consider because the very high DVD disc capacity has been achieved not just by using a narrower channel (which meant that the laser moved from a long wavelength red light to a shorter wavelength blue light), but also there have been techniques included to write information at two different depths in the disc storage layer, and they also have double-sided discs. For music, the extra capacity can allow more music to be recorded and/or there are more channels for the music, so that, at least in principle, using a more sophisticated player one could adjust the way the channels are combined during the playback. These are really exciting options, but they come at a price, not merely the price of more advanced playback equipment, but with facilities that may result in developments of these discs to be only played on new equipment. The future systems may not be able to play older DVD and CD discs. It is of course the normal pattern of improved technologies being developed in ways which unfortunately makes the older systems incompatible. My prediction is that, in 20 years, the current music CDs will be relegated to devotees, who have maintained current generation electronics, and they will be viewed in the same way that current CD users consider aficionados of vinyl discs.

Experimentation with Tape for Popular Music

Mass market recordings on magnetic tape went through several development stages and had value for a mass market as they were small and portable, so they could be carried around or played in a car audio system. In large versions they were, and are, used in many recording studios and are still a primary storage medium for data recorded by particle physicists. Less obvious is how important they were in the development of pop and rock style music. I was fascinated by a detailed account by David Byrne (*How Music Works*) who commenced his career when such tape recording was emerging. It was typical to sequentially record different tracks on top of each other, so in effect the same player could perform all of them. Certainly not easy, as any change in pace (or tuning) would mean the need to start again. Tape was valuable in editing, as one could cut out a poor section, and splice in a repair with sticky tape on top of the magnetic material. Back in the 1960s a craving for new sounds and experimentation included such bizarre approaches as chopping up a recording and reassembling a tape from the fragments, not necessarily in the original order, or even playing in the same direction! Not surprisingly this produced highly unusual effects, and for instruments being played backwards, total confusion (or novelty) for the listener as to what instrument was being played. As mentioned earlier, reversing the starting transients is totally confusing for our brain which is trying to identify instruments and sounds.

Such editing techniques further meant that the output was not defined by normal instrumental playing, but could include recordings of mechanical events, street noise, bird song, etc. The approach is often used in the popular music scene as it is now common in background music and sound effects for films and TV. I have also heard 'classical' music compositions where the composer has intermingled music with recordings of bird songs. The concept is certainly not new, as in the 18th and 19th centuries composers such as Haydn and Beethoven tried to include bird song via both normal instruments and children's musical toys. Vaughan William's violin music of a Lark Ascending offers a 20th century example.

Advent of Compact Disc Recording

The CD systems not only benefit from being smaller than vinyl records, but they have a greater storage capacity on a single side. This is a good marketing point for all types of music and, while not serious for pop music, it has particular value for long classical symphonies. The 25 to 30 minute format per vinyl side, meant that it was difficult (often impossible) but essential to arrange the breaks between vinyl discs so that they came at musically logical points. For example, in a really popular classical work such as Beethoven's 9th symphony, one vinyl side could accommodate the first movement and part of the second. There then had to be a turn-over to complete the second movement and add the choral final move-ment. Worse was that the entire symphony was still too long for one disc, unless it was a rapid performance. Indeed, this symphony has par-ticular importance for the development and success of the modern CD as the original design stage in 1981 was for smaller discs than we now use. However, it is claimed that Mrs Morita, the wife of the Japanese leader of the design team, insisted that the size be increased to accommodate the 9th. Thanks to her, we now have a CD with about 80 minutes capacity. In design terms it only increased the diameter by about half a centimetre, but this was nearly 12 per cent in recording time (i.e. about an extra 8 minutes).

This 80 minute capacity can cope with even the slowest renditions of the 9th, but Beethoven is not the only beneficiary. Composers such as Bruckner and Mahler have symphonies that were too long for the original design but fit within the 80 minute version. Bruckner number five clocks in around 72 minutes, but Mahler symphonies 2, 5, 6, 8 and 10 would all have been problematic on the smaller disc, with overall times between 72 to nearly 79 minutes. Mahler symphony number three is still a two-disc composition, as the six movements require nearly 94 minutes (and a lot of audience concentration—but it is enjoyable).

In terms of technology, there is a totally different approach between vinyl and CD recording. The vinyl is just a continuation of the 19th cen-tury approach where the sound intensity makes impressions in the wall of the signal channel. It is thus a direct analogue of the pressure pattern of the sound wave. By contrast, the CD approach was to reprocess the sound waves into a digital number of 'on or off' pulses (i.e. a zero or one) at each sampling point. If sampling is made at a sufficiently high data rate then they store all the same information as was on the analogue source. In terms of the engineering, one just has to choose a fast enough sam-pling rate of the original sound wave pattern. Intuitively we can guess that if the wave is oscillating at 500 Hz, then we need to be looking at the shape of the wave pattern, and recording the intensity electronically, at least as often as 500 times per second. The more detailed engineering analysis supports this guess, and says that we need to sample at a rate which is at least double the highest frequencies that are to be reproduced. Therefore, for music we need to sample at a frequency of say twice the sound range we can hear (and hope to reproduce).

As young humans the highest notes we can hear are around 20 kHz, and the CD sampling rate has been typically 44.1 kHz. This is both electronically simple and musically acceptable. Indeed, since much of the classical music audience is of a more mature vintage, the upper frequency limit of an elderly audience may be not much more than 12 kHz. The sound delivered from digital recordings is at least as good as many listeners might hear in a live concert. The real delivery problem is not the CD disc but the quality of the subsequent amplifier, speakers, and room acoustics.

A computer store and processor use exactly this digital type format. Digital recording of music on magnetic tape systems is encoded by reversing the direction of the magnetic field on the tape. By contrast, for the CD, the zero and one levels are optically written by having patches of different reflectivity for a small laser beam. The changes are normally from having physical pits in the surface of the spiral track, but it is also possible to give two reflection levels by changing the reflection or absorption in spots along the path. Both vinyl and CDs have a spiral track, but the CD version is designed to read outwards from the centre.

The key bonus of the laser reflection is that the intensity measurement is made without any physical contact with the surface, an excellent improvement as there is no wear and, with some skilled data processing, no dust noise.

In terms of physical stability the CD is more rugged than the vinyl disc, and commercially written music discs will probably last 20 years. Most users of CD storage for their computer operations will have experienced an occasional disc failure. Heavily played CDs, as from a music library, will typically show problems after maybe 5 years. As the CD has only existed in general use for say 35 years, then we can probably assume that these problems will be much more evident from our music collection over the next 20 years. Home copying and storage discs have a shorter lifetime than the commercial ones because they use different chemical formulations on the disc surface to encode the reflectivity changes. This reduces their long-term stability. Failures include physical, chemical, and biological damage. Chemical attack can even be caused by the solvents used in the plastic packaging and storage boxes. Less obvious to European users is that there are many bacteria and bugs that think the plastic surfaces are very tasty. I have a friend who has been studying CD damage from more tropical regions where bugs and bacteria have destroyed patches of the surface. He has many examples.

The format for electronic storage on the CD is potentially an additional problem because CD formats are not sacrosanct, and electronic systems to read them can change in the read format, and/or for computer systems in the size or packing density of the information. The more mature computer users will recognize this problem as they will have seen pre-CD storage discs in the 'floppy' formats of 8, 5.25, and 3.5 inch style, which steadily increased in capacity, but are no longer readable by a modern computer, both because such reading drives are not included and because the relevant software has vanished for most purposes. Amazingly this format rejection is not universal because within the last 15 years NASA have been searching for such discs, as they were used in space flight control! Basically, they had a tried and tested working system and did not wish to risk modernization. Their in-flight computers were also less powerful than a modern phone. This is a thought that astronauts probably do not wish to dwell on.

Changes in the software to read CDs that we have written can mean that one's best family photos could become unreadable. From all the historical precedents of information storage and retrieval this is going to happen (indeed many people say it already has). The only solution is to recopy such information onto newer format systems. This may not be possible for our CD music collection, so we assume, and hope, that the same types of CD player will be maintained. If not, then we are in the pianola or vinyl situation, where we have music encoded on paper rolls, magnetic tape, or vinyl, but have no easy way to play it. The loss of future equipment to play current CDs is inevitable as electronics companies try to develop newer and more competitive products. Just as gramophone records have been remastered onto CD formats, we can expect that much of the more popular CD music will appear on the new 'super whatever' system of a future technology but it will mean repurchasing them. More obscure items or limited editions will be lost.

In discussing information and musical loss I am not being pessimistic but instead am using historical examples to recognize that no recording method has survived indefinitely, and for electronics the playing field is extremely dynamic. Indeed, most laptop computers now exist without CD drives.

Rather than a CD system, which has moving parts, one can consider electronic storage on flash memories (i.e. memory sticks), MP3, or MP4 players, or computer-style memories with terabytes of storage capacity.

Memory sticks are small (and easily lost) and can be corrupted as well as becoming obsolete in terms of the format for newer computer systems. (Note the mark 1 versions are not readable by modern equipment.) MP3 systems are acceptable for some types of listening, but they use compression techniques in order to pack more information into them. The compression degrades classical music, and so there is a trade-off between capacity and sound quality. A degraded sound may be fine for low concentration listening (i.e. background music) but it is very poor for high quality reproduction systems.

I am sure this is a section which will need an update within a few years, especially if the storage capacity is included without loss of music quality or, even more interestingly, if other features are included such as the ability to change the balance of sound between the recording channels. A feature that I would personally like to have available is the ability to alter the dynamic balance between soloists and accompanists or orchestral backing, since in many cases the recording studios over-emphasize the high profile soloists. The reverse is also apparent on some discs with soloists hidden in the total sound. This is worse than changing seats in a concert hall. It would be a desirable nice option. I have heard opera recordings where the soloists are excellent, but the acoustics make them appear to be in the far distance, and I would like to move from my virtual and distant gallery seat down to the stalls. These may seem futuristic options but they are not even beyond current technology. Therefore, I predict that they will emerge as later options, when the market forces need to generate new musical toys that make old equipment obsolete.

Local or Cloud Storage

We have a misguided hope that the future may be better and that some new advance will solve all the problems. One concept that is currently fashionable is for centrally stored, and backed up, electronic information. In principle, one's music collection could be accessible from any place with suitable equipment. Storage of course depends on payments, so failure to pay, or death, could mean that all such records are lost. This is distinctly bad news for future historians or biographers where all correspondence has vanished into electronic formats. It also implies that we would need to have the ability to access the store, which may not even be in the same country. Therefore, there are communication links of satellites or optical fibre cables within the access process. Despite their current fantastic performance these cannot be guaranteed. Potential problems are data jams on overloaded information highways; political interference (i.e. not just monitoring of internet traffic, but actively blocking or changing it); accidental or intentional destruction of the routing (e.g. if one looks at the fibre optic route maps around the world it is clear that a seismic event or a terrorist or politically motivated attack could seriously destroy major highways). Finally, there are magnetic storms on the sun that have interfered with electrical communications on earth (even the simple US telegraph in the 19th century). The major magnetic storms have occurred randomly on a timescale of centuries but there has not been a large solar event for a long time. Therefore, there is a steeply rising probability that a big solar outburst will happen, and this will seriously disrupt (or destroy) both satellite and ground-based electronic systems.

All such examples just demonstrate that no matter what information we attempt to store, whether art, literature, or music, the storage system will have a limited lifetime, as will our ability to understand the information that we can access. The positive view is that this is exactly our own personal problem with memory, and yet we survive despite information and memory loss.

Where and How Do We Listen?

Having carefully performed the music, edited and processed the tapes to remove undesirable features, and then adjusted the tonal and frequency balance to the best of the engineers' ability we feed the CD into our home player. If we are lucky this will be a good match to the electronically prepared CD and the final sound output. It should have high frequency fidelity, and the dynamics intended by the musicians plus, with two signal channels, a sense of stereo. We now have two options. The first is to listen via high quality headphones and the second is to project the sound via a speaker system. Headphones have several advantages if only one person plans to listen to the music (parallel headphones are feasible for more people but this is less common). The headsets certainly guarantee that stereo sound is offered to each ear. It will not be perfect because the sound will be projected rather too much into the head and we will lack some of the multiple reflections and time delays that we hear in a live event. There are also a few phase issues in the digitally processed sound that come from the noise reduction techniques. Most of us will not easily recognize what is odd, but certainly the quality may be slightly impaired and may even have an occasional flutter sound. But a real bonus is that it mostly isolates us from the acoustics of the room and the extraneous noises that exist in a home.

I have also discussed with people using hearing aids that using headphones is often inadequate as they are not personalized to the individual correction frequency response of each ear. Normally they cannot be used over hearing aids. Electronically it would be simple to make an interface between source and headphone that could have a personal pre-programmed frequency response. The potential market is large, especially for classical music where perhaps 50 per cent of the listeners are approaching the need for hearing aids. Less obvious is that this interface could span a greater frequency range than is included in standard hearing aids which typically stop by 6 kHz (i.e. the maximum range of human sounds produced during speech). The interface correction package need not have this limitation. Overall, the result would be a delivered sound to a person with partial hearing problems that would be better than from the hearing aid. The potential market is large, so I hope this is appreciated by the electronic industry.

The alternative of speaker projection means we do not need headphones but the top-quality sound will generally be available in a limited area of the room. We are also totally dependent on the acoustics of the room, which may well have many soft furnishings, carpets, curtains, etc. that deaden the sound. If we find this too irritating then we can find a different room that is dedicated to listening to music. Although not currently relevant for standard CD packaging, many home movie systems (and some TV) broadcast via not just two speakers but with perhaps four, to include some directionality, some surround sound, and a floor-mounted speaker to enhance bass notes. For cinema, this surround effect is impressive and in principle one could package music which has been engineered to exploit the multi-speaker dedicated room delivery. Cinemas led the way in the use of stereo, and then in surround sound in theatres, so it may well be the forerunner of home surround music. The technique is feasible as the DVD format can store roughly 25 times the amount of data as on a CD, so extra music channels could be accommodated. However, the disc would not be suited to a standard CD player.

Again, it is an option with some advantages and it would require us to purchase more equipment, so I assume it will be developed and marketed.

Fortunately, the brain is highly efficient at dealing with limited information, and so to some extent we ignore the problems of room acoustics, even though we have the ability to hear better sounds in a concert hall.

There is an opinion that the market of CD recordings is doomed as it is now possible to pack more data on an MP3 or MP4 type devices, and so the CD can be replaced with these small units with greater capacity. Superb marketing hype, but musically it is false. The MP3 packing was designed to work by removing as much information as possible from the recording. An apparently justifiably view would be that frequencies above our hearing range could be discarded, and since most audiences were not fully concentrating, while listening via headphones and/or simultaneously doing something else, even more frequencies could be rejected. Classical music may well have a high percentage of older people whose hearing has degraded, so this is another reason to reject the original music at high, or low, frequencies. The fallacy is that, although we do not *consciously* recognize high frequency notes above say 15–20 kHz, the brain apparently starts signal processing on far shorter timescales. I have seen reports that this is equivalent to signal changes at 40 kHz. At a purely personal level, I have never heard successful reproductions of classical music via any MP3 system.

Many people have commented that modern live performances of various types of music are now consciously distorted, so they sound like the modern CD or broadcast versions because this is what the public wish (or expect) to hear. So in effect, little has changed in 150 years. Just as one example of the recording dominating the performance expectations, many bands, groups, and classical performers will be far less strict in exact timing when they are playing live, as this is part of the excitement and presence of an interactive communication with an audience. It is true whether in a nightclub, pub, gig, or classical concert. If the CD were to sound that flexible in the timing and dynamics, the listeners would not necessarily appreciate it. This feedback implies that the CD version then becomes the expectation at a live event. Technology is driving the music, not the other way around, and we need to resist this trend. As I mentioned, in a studio recording, patterns have changed so that in some music all the individuals are effectively isolated in echo-free environments. Players say this destroys the performance, and the subsequent electronic 'adjustments' can never regain the originality of the united group.

The Future

Predicting the way electronics, computing, and listening to music will develop in the future is all too easy. It is only in hindsight that we will show how far we were wrong. Based on the last 50 years electronics (and computing) have advanced in ways and at speeds that would have been considered totally impossible 50 years ago. The experience with all types of storage media is that they have a very limited operating life before they are superseded and, since CDs have already been in use since the 1980s, it would not be surprising if they were displaced by new storage media, as sketched in Figure 11.1. CDs came, dominated, and then faded within two decades. The really big advantage of such replacements could be higher storage capacity so that, with suitable electronic equipment (i.e. a big market opportunity), we might be able to modify the sounds we hear to suit our tastes and listening environments. With more tracks, we may perhaps open a route to personally alter the channels that balance soloist

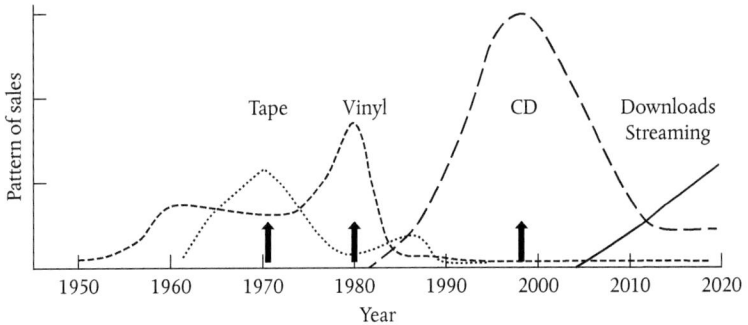

Figure 11.1 A very approximate indication of the market for different types of recordings over the last 70 years. While the 78 rpm systems have vanished, vinyl records are having a small resurgence and CDs maintain a steady market for classical music. Projections for downloads are speculative but streaming is increasing rapidly. Each past technology has dominated for around 20 years.

and orchestra, members of a quartet, or other features that at the moment are merely part of the overall package.

Equipment costs seem to have been remarkably less critical than one might imagine. In 1982, the original CD players were selling at around $730 (i.e. equivalent to roughly $2000 in modern terms). But by 1988, CD sales exceeded vinyl and then overtook tape by 1991. A key factor with such electronics is that the advances are matched by falling equipment costs.

The reason for the rapid drop in CD sales since 1988, as sketched on Figure 11.1, is difficult to quantify in terms of where the new plateau level of the sales is. It is the result of a vast number of young people preferring not to own a copy of the music but always to listen to it live off the internet, and/or to download specific tracks instead of buying an album.

The download and streaming route is still in its infancy for classical music, so here I think the demand for CD and vinyl is more secure, and that classical style music will persist for a very long time; items we now think are avant-garde will seem more acceptable and mainstream; new instruments and the synthesizers will continue to develop and our generation will claim that the new sounds are not as good as they were when we were young. That is life.

Normally I am fairly optimistic about the future of music, but I recently watched a TV programme of a rising pop group in which the lead member said that there is now no need to be skilled at performance, as all the sounds can be composed by, or prepared on, a computer and synthesizer. He saw the live event as just stage actions to the ready-made 'music'. For me, this is not just a total failure to appreciate the role of music (both classical and popular), but it totally undermines musicianship across the board. Part of the excitement is the essence of freshness, originality, and the adrenalin that is the pleasure of hearing real live performance—plus the danger of making mistakes and the pleasure of hearing spontaneity. Equally, I have heard computer experts claim that artificial intelligence using self-learning computer algorithms will be able to compose music in any particular style, and then will develop their own tunes and displace human-written music. The AI items I have heard so far were dull, boring, and lacked any musical originality. Perhaps the market will be for computers that listen to AI music. Once we remove humans from writing, playing, and enjoying music then our place on the planet is doomed.

Downloads and Streaming—Will They Change What We Listen to?

Predicting the future is clearly harder than reviewing the past, but the immense decline in CD sales is squarely the result of a different way of accessing music. This is a radical change in the way music is marketed and distributed. It has severe implications and immediate impact on the music retail, distribution, and recording companies, and it has highly unpredictable consequences for the funding of producing all the tunes we love. For example, the retailer HMV has recently gone into administration for the second time, with streaming cited as a contributory factor. The situation has arisen from advances in electronic communication technologies, which at first sight may seem highly desirable, but have uncertain consequences both for popular and classical performers.

The problem is worth considering as, particularly for the smaller area devoted to classical and jazz music, it could be the death knell for large sections of such music-making. As a scientist I am well aware that innovation and technological advances frequently have unexpected and negative consequences. Indeed, I have written precisely about this topic (*The Dark Side of Technology*). For music, the truly negative features are in the near future, so it is essential to discuss them at this early stage to see how they may be avoided or mitigated and so inhibit the collapse of support for the recording industry. Classical music is particularly vulnerable because for many companies it can be both expensive and have a limited market. Also, many works are very long relative to a typical pop music item.

As was very obvious from Figure 11.1, the sales and total domination of the music market by CDs went from nothing to a peak within about 15 years but then plummeted as purchasing patterns started to change to download popular music as single tracks, rather than as albums. Initially, downloads were often pirated and free. This has set the pattern and expectation that access to music should be at minimal cost. Downloads are still effectively free, as a small monthly payment will allow unlimited access, or initial subscriptions are zero with unlimited downloads for a period of time. Downloads must be stored somewhere (computer, cloud, or mobile phone). This led to the alternative, which has become more fashionable, of not keeping the items but rather to continuously listen by direct streaming from a central source. This generates a major demand on the transmission technology. The bonus is that the music can be heard via a mobile device while travelling, or at home via local transmission devices using say Bluetooth. For background music quality, this is adequate because Bluetooth technology is a low-power wireless device to send audio signals from mobile phones, tablets, or laptops to headphones or speaker docks. Many streaming sources also suppress the quality in order to cope with the volume of demand. This is accepted because sound quality on earlier phone earpieces is poor, even for speech. Later style large headsets are somewhat better.

In the case of popular music, in all its rapidly changing guises, people like to listen to short items from a performer, and often prefer this to buying an entire album. It is therefore better (and cheaper) to select a particular tune and artist, hear it, and move on. This means album sales are down, and the listener may well have missed many other items that would have been equally enjoyable. Compared with CD album sales, the income to the industry has reduced by around 50 per cent. The Official UK Charts offers a useful guide to the changing trends in distribution and sales. For example, in one week in December 2018 the number 1 Single—(Thank U, Next by Ariana Grande), had total sales 55,405, of which downloads were 4073 and streams (sales-equivalent) 51,332. This is for an item that had already been an incredibly positive piece at number 1 for 4 weeks and an accumulated total sale near 300,000. By contrast, the total album number 1 sales (Odyssey by Take That) were 105,721, of which 99,034 were 'physical', downloads were 4922 and streams (sales-equivalent) were 1765.

The 'instant music wherever you are' idea seem fine as long as high-quality music is not needed, since streaming is very costly in terms of data (and signal bandwidth) and so for most music the work is compressed. This means that it loses sound quality both in the range of frequencies and the range of sound levels. This is acceptable for continuous background and music that has only been heard by electronic transmission. To a high-grade music aficionado, the sound is often far below an acceptable level. This applies equally to all music (i.e. from classical to jazz, etc.). Realistically, streaming is targeting an enormous mass market of popular music where the demand for quality of the sound is often very modest. To add some perspective on the audience, a survey in the USA in 2017 claimed that the average time spent listening to such streamed music was at least 32 hours per week, and this number is now thought to have risen. It is clear that people are listening while doing all other aspects of travel, work, and relaxation, and for such a continuous background feed it is highly unlikely that they are able to be very critical. Equally, it means that all other aspects of what they are doing are similarly undermined. Related economic surveys of streaming found that some 75 per cent of Americans used streaming every week, and an equal percentage said they would only use free access sources, whereas in the past they had bought vinyl and CD formats.

The typical five minutes of pop song is delivered at an incredibly low cost to the listener and inevitably this means that the payback to the artists is equally minimal, unless there are a phenomenal number of listeners to the track. At this point, intuition is useless as we may have no idea of how many times a particular song is streamed, nor how much money it may pay to the artist. Numbers that I have seen quoted in 2018 are around $0.007 per stream to the holder of the music rights, where 'holder' includes the record label company, producers, artists, and songwriter! For most people, this is effectively just a token and not a successful career. In hard numbers, a million plays per month might generate the same as the US minimum wage of $1260 (in 2015). Media stories and the Top Ten lists often report the exceptional successes, and for a song called 'Shake it Off', Taylor Swift had some 46 million streams and she is estimated to have received a total payment in excess of £300,000. Note however that by number 10 on the list the sales are down to 10 per cent of the leader, and often continue to fall at this steep rate with lower-placed items. Here, the downloads are typically just 10 per cent of sales, and streaming will exceed the purchases.

The potential for instant fame and financial success attracts amateur creativity. It can be fun, but shortcuts to success by recording at home, and marketing directly via YouTube actually pay very little, and the numbers can be as bad as 4 million plays per month to reach the minimum wage threshold. Here there is a further problem, since as more people take this route, then there will be fewer viewers per item. From the listener/viewer perspective, there may be an even more dramatic fall in quality of hearing people who have zero training or professional advice.

The power of computer algorithms to register all the downloads that are made means that many people now allow software to broadcast to them music that fits their typical listening pattern. Apparently, many listeners will only listen to playlists of their own musical taste and not attempt to choose any item themselves. This is a self-reinforcing narrowing of one's musical perspective. It also exists in many radio stations, such as Classic FM, where the same fragments from a very narrow range of compositions are played. It inhibits listening to other composers or other compositions and so erodes one's musical development. I have already mentioned that there is a wealth of really great music that we rarely hear unless we make a special effort, from composers who do not find favour with broadcasters and/or recording studios. These changes in distribution and self-selecting algorithms that curate our tastes are pernicious, as they will destroy the wealth of musical heritage, and determine an ever-narrowing future musical selection available to us.

At the current time, the vast streaming market is aimed at popular music and to call it up one needs only to quote the name of the song and the artist. This is relatively successful because this may give a unique identification, since topical items will be limited to perhaps just one recording of the item. By contrast, the present situation for classical music is basically a disaster in terms of streaming, as far too little information is required for the selection. Recordings over the last 75 years on vinyl and CDs may contain hundreds of examples of the more popular classical works, from different performers and orchestras. Indeed, one can easily find how recording techniques and musical interpretations have evolved by comparing them (often from the same conductor or performer). Streaming has not yet hit the classical end of the market, but improved identification and selection means that it will appear, and unless there is a major change in streaming quality, it may only have third-rate sound. As this happens, the higher quality from vinyl and CDs will no longer be available. At the same time, there will be no financial incentive for recording companies to produce minority items, or truly expensive ones such as operas. Perhaps live recording will be stored and these may become material for subsequent streaming. Examples I have seen cited are that requesting a Beethoven 5th symphony may give one (not necessarily the one asked for but number 3, as this was opus 55) or a totally different composer with the same number symphony.

An extreme pessimistic opinion could be that, while technology brought us recordings that spread all types of music across the world, the new variants may have totally the opposite effect and just reduce the broadcast to poor-grade sound, unknown artists, and minimal funding for the performers and composers. Similarly, the search algorithms and playlists may produce an even smaller set of available examples. A modified music industry will survive, but we, the listeners, may have less choice. Many companies are actively writing playlists that define content, and they offer only a limited range of artists. They see a financial future in such playlists and are encouraging the use of streaming. 'Classical' broadcasts of playlists often mix in a wider range of music, but again do not necessarily define which artists are in use for the source material. For items such as a Beethoven symphony there are hundreds of recordings, and the listener will not know which is used.

Precisely because the financing is now precarious and there are efforts to include longer classical items, the charging patterns may change to cost per second rather than per item. This would raise the classic charge as concertos and symphonies etc. are often 10 to 60 minutes long, rather than the standard 5 minute pop item. Income from streaming is around a mere 0.3 per cent from classical music and just 1 per cent from classic CDs. Overall it means development of classical music distribution is very low on the priority list of the companies.

A more positive view from Dr Clemens Trautmann, the president of the classical label Deutsche Grammmophon, is to embrace the opportunities of streaming to make it widely accessible with subscriptions, and also expand into videos of more classical music. This includes their first video entry into opera (Romeo and Juliet by Gounod). A great deal has been written about the situation and the possible future, survival, or disappearance of different performers and recordings. So in my final list of references I cite the extensive summaries and blogs by Mark Mulligan (from MIDiA) as he has a very balanced view of the complexity and uncertainty in the industry.

CHAPTER 12

THE VOICE

The Intrinsic Music of Humanity

In the preceding chapters I have rapidly tracked how music was constrained by political and religious forces. In parallel, there was considerable technological input in terms of improvements to existing
instruments, inventions of new ones, and the dominance of keyboards,
which forced a shift in tuning. Overall, these factors altered our perception of fashionable composition and performance, and they have led to
new expectations and availability for us to hear worldwide music from
first-class performers. In many ways the improvements or changes to
instruments and the addition of new ones were the result of a wide range
of different technologies. Their musical impact was dramatic because it
drove fresh directions in composition and style. This was a one-way
path, and there is no turning back. While such changes are real, they
totally overlook our oldest and greatest instrument—the human voice.

Many creatures, of all sizes, from whales to birds, and even some
insects, intuitively sing. With all animals it covers our emotions from
happiness to sadness and it is frequently involved with seduction. The
birdsong spring choruses are fine examples of the mating instinct. The
only difference between the birds and ourselves is that birds tend to have
a fixed repertoire and are less likely to have impromptu variations.
Whales will happily learn new songs from other whales but again it is a
group activity and the new songs are repeated. There are no additional
developments, modulations, variations, or virtuoso cadenzas that we
can distinguish. This is not a problem of brain size or ability to communicate. Mosquitos do not come to mind as amorous creatures but male
mosquitos lower their frequency (from wing beats) to match a favoured
female sound.

Humans are equally interested in mating, and over the years there
must have been millions of love songs created (after all there are now

over 8 billion people). We can only guess when music for entertainment, storytelling, and religion first developed but it was certainly many millennia ago. Our difference from most other creatures is that we can be flexible with the tunes, the words associated with them, and the style of delivery. This is just as true for group singing activities, whether religious, social choirs, or patriotic, and military music. Any focus on the pop culture reveals that we can learn new songs at a rapid rate, as trends, groups, and their musical styles fluctuate on very short timescales. People remember the tunes and the singers so in terms of repertoire the general public are deeply immersed in songs and are able to vary them on an individual basis, at will.

There are also the professional singers, ranging from pop culture to grand opera. The only difference for the professionals is that their performances must meet higher standards than us, the general public. This requires some inherent ability, considerable dedication, and training. This is no different from any other field of endeavour, whether science or sport.

If the voice is musically so important, then why has there not been a huge technological input at source? However, humans are the source, so this is not feasible as we are mostly running on the genetic inheritance from the last 10,000 years for speech, language, and vocal music. In technological terms we cannot make improvements, except those that contribute to our voice production, learnt via lessons, training, and repetition. In terms of mass market changes by additions of new technologies, we have no idea how to modify either the voice or our sound production, though I will ignore surgery or sound synthesizers for those with speech difficulties. For the majority, what is sung—especially in private for our own pleasure—is with an instinctive voice action, and there is no science that can make improvements at the initial voice production stage, whether for true professionals or live performances. Technology can however be helpful in hearing our efforts during training, and professionals will spend many hours practising so this is valuable science.

Practice is not an insignificant investment of time if we wish to be a professional musician. For instrumentalists and singers the guideline is that it takes around 10,000 hours of practice to reach a standard that would be acceptable for entry into a musical academy. For a classical music singer it is typically a ten-year commitment path to reach a marketable level with a chance of fame and wealth. Popular music singers will go public far sooner and are able to do so as their voice production will be more similar to that of speech, plus they heavily exploit electronic recording and sound generation. Starting with a language that is musical and a culture where singing is popular are both helpful and desirable. From a UK perspective this means that there may be some advantage in having a Welsh background. On a global view others will claim that a language such as Italian is a good starting point. Tonal languages are also an excellent intuitive input, so it is no surprise that we now hear singers from say Manchuria performing the classical Western repertoire.

The conclusion is that while the voice is an instrument, we can only benefit from technology in training and subsequent sound processing, in complete contrast with instruments such as the piano, clarinet, etc. which have been engineered into new variants. This is not irrelevant because there are all the peripheral factors of accompanying instruments, recordings, broadcasts, and, at least in the pop world, inclusion of 'autotune'. For those unfamiliar with this technique, the computer software takes the notes delivered to the microphone by a singer. If the note drifts off the intended one, the frequency is electronically 'corrected', before it is played to the audience. Sad to say, it is now an essential technology in many recording studios. As with all such electronic technologies, it can be used to alter not just the frequency but also the attack, how long the note is sustained, and the tone quality as one sings from one note to the next. This is valuable for a poor performer, but it totally destroys the subtlety of a good singer, who will actively alter pitch for musical impact. In the popular music market, such signal processing can reprocess the performance to add different sound quality (timbre), and take a thin voice, and make it seem much more powerful, and rich in texture. Many microphone-based singers are very unimpressive if heard solo, without their electronics. At the opposite extreme, opera singers will not merely resound across a concert hall but do so in the presence of a large orchestra. Wagner may provide extreme examples but the non-electronic approach is normal. There is no value judgement in this comparison because both types of singer appeal to different mass markets, and both can be highly successful.

Popular music used to include many crooners who had benefitted from classical training, as well as performing with microphone delivery.

A prime example was Frank Sinatra, whose career blossomed after having singing lessons from a member of the New York Metropolitan opera. He moved from just being successful into the superstar range.

Separation of Song, Dialect, and Language

Voice and speech are clearly related, as they are normally language based and in both cases we have an incredible ability to vary the meaning by tonal control, as well as, pitch, pace, and volume. I have carefully said 'related', not 'identical'. People who stammer can have excellent unbroken diction when singing, I know many people who speak with strong accents, or when using English as a foreign language, persistently mispronounce our somewhat erratically spell words. For them it is not unusual to be able to sing smoothly in an impeccable accent. In old age, dementias can undermine conversation but the ability to remember songs and sing will persist. Voice is thus using brain power in ways that we only partially understand and under-exploit. The power of our voice control is that, if we choose, we can reproduce songs in dialect, or foreign languages that we do speak.

Music teaching in schools is important, but at times of financial pressure it can incorrectly be deemed to be an unnecessary expense, and recent data (2017) indicate that in UK state schools, music staff has dropped by around 40 per cent during the preceding five years. The style of music that is taught has also changed. In particular, classical music has been considered somewhat elitist, and so emphasis has shifted to more popular styles. To me this is unfortunate as it is failing to use a less obvious feature of music, which is highly effective in the teaching of foreign languages. An obvious opportunity exists in the very early phase of such teaching. Singing in a foreign language is feasible and can easily be taught because it bypasses all our normal biased reactions against learning words with unfamiliar pronunciation. I therefore feel that it should be used as a *precursor* before speech, vocabulary, and grammar. It is essential that this control of accent should precede reading, as once we know we sound like the real thing (i.e. in French, Spanish, Mandarin, etc.) then we will have immense confidence and learn far more rapidly. Trying to add and correct accent at a later stage is extremely difficult, as the faults are firmly fixed, and we do not (or cannot, and perhaps will not) recognize them.

A personal note is that my poor Anglicized French accent was transformed when I started learning songs in French. Similarly, when working with Spanish colleagues I refused to read any Spanish, as this would have meant I imposed my English accent on the words. Instead, I listened to people and videos and tried to duplicate spoken Spanish. It helped, because although my vocabulary is limited the accent is acceptable and I am understood and have many Spanish friends.

Attempting more challenging languages can be worthwhile, even for very small specific occasions. When in Beijing, I carefully learnt a sentence in Mandarin to open a public lecture. This produced a round of applause, and a receptive audience for the rest of my talk. Tonal languages such as Mandarin are musically interesting because the skill needed to use the same word—but with changes in the pitch to provide different meanings—requires a very acute ear and voice control. One consequence is that such tonal speakers tend to have a far better sense of pitch than say Europeans. For example, they can automatically speak the same word on the same note. The Mandarin speech pattern includes saying the same word in four different ways: speaking at a fixed frequency, going down in pitch, rising, and going down and then rising again. Four different words from one word is confusing until one realizes the very significant importance of the changes in pitch. The written characters for the four words are of course quite different. The four examples in Table 12.1 underline the skill needed for both speech and calligraphy. They also reveal the difficulties for foreigners in deciding what is the meaning of the sentence (e.g. male and little girl are quite different, but are the same word).

Before being critical of a language that uses one word for several meanings, we should realize that spoken English does this all the time,

Table 12.1 Multiple meanings of some Chinese sounds depending on pronunciation

'Ba' can mean:	eight 八, father 爸, target 靶, pull 拔
'nan' can mean:	male 男, difficult 难, south 南, little girl 囡
'huo' can mean:	fire 火, alive 活, obtain 获, confused 惑
'shui' can mean:	water 水, sleep 睡, tax 税, who 谁

but we are far inferior because often there is absolutely no way aurally to distinguish between the alternative meanings. With Mandarin, the written versions will differ, but in English the written word is often identical for all meanings. Examples are numerous. Probably one of our most prolific words with both identical sound and spelling is 'set'. My dictionary offers a total of 58 meanings, plus two more for sett! There are many more examples with at least 20 alternatives.

Voice Training and Singing Teachers

The internet has opened access to many guides, training, and exercises on how to sing in different styles. The guidance is often free, and frequently helpful. Nevertheless, such small package courses can never truly address specific problems, and in many cases the text, and words, that explain how to be a singing success are difficult to comprehend. Books and articles are numerous. In the background for this chapter I have read many articles and often feel that the writers are probably good competent teachers, but rarely do I gain insights that immediately strike a chord (well, a solitary note for a singer). In part, this is because there may have been several hundred years of singing teachers, but only in the last decade or so has there been enough medical data to recognize what is happening in the highly complex mechanisms of our voice production and control. Visual inspection, with tubes inserted into the throat, or magnetic resonance imaging (MRI), offer unprecedented examples. This is helpful for the medical profession to use hindsight as to what is happening as we make different sounds. The overall processes are not just extremely complex but are subtle and almost totally specific to each person that has been studied. I am fortunate in having an excellent singing teacher who has a very good grasp of the physiology, of action and of the effects of reshaping the various components in the throat and mouth. My understanding is at a far lower level but I can follow the broad description. This means that when I make some changes in the way I try to produce sounds, I can then hear different tone quality. Although there are new tones, I am rarely conscious of which muscles the brain is adjusting. Practice leads to subconscious control of the correct muscles. This is a normal response because if I pick up something, I do not actively need to consider my hand actions. A key step is that when my teacher sings, I find it easy to copy (we are both baritones), and this is far more effective than thinking about the mechanics of the muscles. I also suspect that having a good teacher who sings in the same frequency range is more effective than is normally stated. This is probably not a comment that is financially publicized by teachers. The advice I am given is to not just to listen to the voice, but also 'listen' to the muscles, and take care with the posture.

Starting singing lessons was instructive in several ways. I incorrectly assumed that having once been a boy soprano in a choir would mean that, as an adult, I would sing as a tenor. Similarly, I assumed that boy treble voices would drop to the baritone/bass range. While this can happen, there seem to be many examples that the boy sopranos are the ones who become baritones and basses, and trebles only move down into the tenor range. One can offer many examples in different genres. Aled Jones was a very good boy soprano and is now a baritone. A more extreme example was Jacques Imbraillo who, as a boy soprano could sing the part of The Queen of the Night (a really high voice part). As an adult, he is a successful classical baritone.

There is a definite consensus that posture and breath control are essential first steps, and these can produce immediate improvements over simple instinctive singing. They can also extend the range of notes that are usable. In my case, I rapidly gained at least a third at the lower end of my range. Less clear cut is that, for men, the way we control the voice breaks down into two or three different zones. The boundaries between them are called break points, or passaggio. For a baritone there are two such zones when initially learning to sing.

Writings on singing technique are very variable in terms of mechanisms and have not just changed with time but continue with strong differences of opinion and of the terms that are used. This is not surprising because there are many muscles involved that control the vocal cords in the larynx, the soft palate, and the shape and position of the tongue and lips, etc. Their actions are interrelated and define sound quality, tone, and power, for example as discussed in the terminology of the Estill method (Steinhauer, Kimberly, Klimek, Mary McDonald, 2017). Not only are the sounds that are produced inevitably related to the physical structure of each singer, but they will also be influenced by the native language (and dialect), plus the ways in which the singer uses the various vowel sounds, and the volume level that is generated. Indeed, as linguistic beginners, we will have no difficulty recognizing that a work is being sung in German, Italian, or Russian, even if we have no idea about the text.

These variables are important in any discussion of the break points. For simplicity, one can sense that on singing progressively higher notes in the normal full voice, the vocal folds stretch and get longer and thinner. Some describe the ranges in terms of where one can feel the notes (not where they are produced). Terms used range through low (felt in the chest), to middle, and high (felt in the head). Classical training aims to make the transitions between the three modes as smooth as possible.

The upper passaggio point may appear to shift between notes, depending on the volume, whether one is ascending or descending, and with the vowel sound that is being sung. Failure is often very obvious, as it sounds as if the voice has cracked between them. In extreme versions it may be deliberate, and the effect is yodelling. For quite different reasons, it can sound similar to the type of voice experienced by some males during puberty, when their speaking oscillates erratically between different pitches and resonances.

The pattern is broadly seen for all singers, and smooth performance needs to cope with these subtle responses of the larynx etc. Training and practice are effective but individuals are still limited by their inherent ability. For the superstars, especially with classical music, there are many challenging arias which have notes or passages that only appear easy from a very few performers. Images of great tenors such as Pavarotti hitting, holding, and sounding wonderful on extreme high notes, fully justify why these are known as the money notes. More careful viewing of the facial expressions from such performances can equally reveal a degree of fear that success is not guaranteed in all cases. Fitness may help and I have read that Pavarotti was once a sports teacher.

The simplified diagram of Figure 12.1 indicates the approximate ranges for different voices. The notes labelled with an 'm' are the ones described as the money notes. They are the ones that only the top performers can hit and hold with control and good tone. They are also the notes that opera composers tend to use to excite an audience, and simultaneously cause fear and panic for many singers. The keyboard offers a clue to the ranges.

On the keyboard A4 is at 440 Hz, and it is the note used for orchestral tuning. Middle C (near the lock on old pianos) is at 261 Hz. It is called C_4, as the musical scale notation starts from a subsonic zero value of C_0, which is down at 18.35 Hz. This differs from the possibly more logical notation

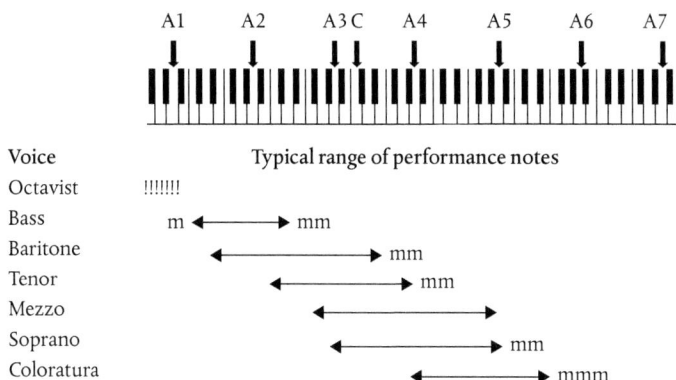

| | A1 | A2 | A3 C | A4 | A5 | A6 | A7 |

Voice	Typical range of performance notes
Octavist	!!!!!!!
Bass	m ←——————→ mm
Baritone	←——————————→ mm
Tenor	←——————————→ mm
Mezzo	←——————————→
Soprano	←——————————→ mm
Coloratura	←——————————→ mmm

Figure 12.1 The approximate range of different voices The 'm' indicates exceptional extreme notes called the money notes.

of starting each octave block alphabetically from an A. In some texts this is not understood, and notes can be misquoted into the wrong octave!

As a simplistic overview a typical non-operatic bass, baritone and tenor, differ by a musical third, and encompass roughly two octaves. Operatic stars emerge if they can exceed these range limits, and particularly if they can produce, hold, have good tone, and are exciting on the highest notes (or the lowest for the bass). Mezzo, soprano and coloratura are similarly spaced by about a third. This is one reason why duets are often written a third apart, as they fall naturally into the voice ranges, and it produces a pleasant sound.

The category of octavist (or oktavist) is primarily Russian (or Slavic), and has developed from the need in Orthodox churches to have deep resonant low notes without any musical instruments. Many oktavists can be described as basso-profondo, and in an opera such singers have good power down to D_2 (73 Hz) or C_2 (65 Hz). Russian ocktavists have been claimed to go almost one octave lower. Judging their tone quality on a CD, radio, or computer link is fruitless as the electronics (including both recording and speaker equipment) can rarely manage such low frequencies (not least because US electronics runs at 60 Hz, and Europe at 50 Hz, so there is a conscious effort to block such low frequency signals).

I have not included voice ranges for falsetto singing, or for castrati such as Farinelli. One may at least relate that he spent his last quarter century in the Spanish court, and in his final decade sang to Philip V, the same four songs every evening. Nothing else! He had already hit the money note category.

Tone Quality and Voice Control

The preceding discussion of break points hints at the difficulty in control of the entire vocal tract, from larynx to soft palate, tongue position, and shapes of the mouth and lips. To demonstrate the sensitivity to all these variables one can benefit from modern recording techniques, and instant frequency analysis running on a tablet. Figure 12.2 displays the spectrum of notes that are generated by a baritone using different positions of the larynx. To some extent, changes can resemble those produced by mouth shape, tongue position, and as a consequence the carrying power. For the same C_3 note (~130 Hz) Figure 12.2 shows the very obvious changes in spectral content when the larynx drops from a high, to middle, to low position.

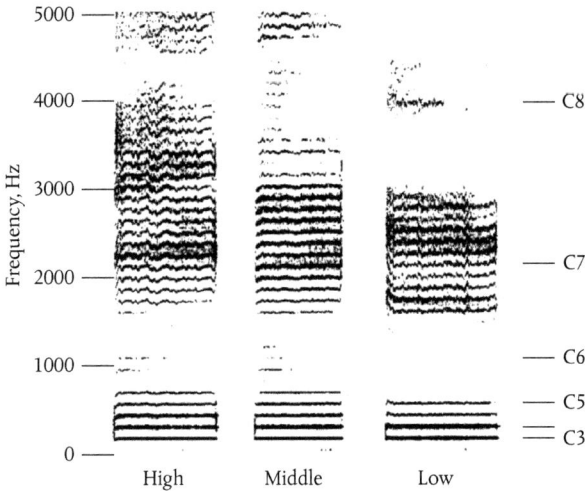

Figure 12.2 A spectrum of a baritone singing 'eye' on C_3 (~130 Hz) for three positions of the larynx ranging from raised, to middle, to lowered. Each example records the data over about 5 seconds. There are many multiple harmonics of the fundamental that are amplified particularly in the 1,600 to 4,000 Hz range. (Many thanks to Stefan Holmström for Figure 12.2)

The perception to the listener is that the high position has a 'lighter' texture to the sound whereas the low larynx has a far more 'macho' tone, a fairly reasonable interpretation because in the low larynx position nearly all the power is in the lower components (i.e. a more male type sound). This immediate analysis system indicates that the singer has hit the fundamental within 100 to 200 milliseconds, and he can sustain it steadily. By contrast, many of the higher harmonics are being dominated and tuned by other parts of the entire vocal tract, and these can fluctuate without disturbing the fundamental. Considering the complexity of the sound generation process this is not that surprising. I have heard a description saying that while the vocal chords are flapping, the surfaces develop a rippling action, and every aspect depends on power, moisture, and the words and vowels that are being sung (plus the previous and future notes). Planning for future notes is frequently essential in knowing how to modify the currently sung word so that one can smoothly progress to the next. As a physicist I can see why textbooks on sound always concentrate on the simplicity of harmonics from a violin string.

A very similar pattern of change appears when singing the same note with an 'ah' sound, but with an open mouth, and with three tongue positions. The base note is unchanged in each case. With a flat tongue, the low frequency components dominate; with a mid-level tongue, there are higher components; and with the tongue raised in the roof of the mouth, a more strident and penetrating sound is achieved. This is desirable in a large hall to gain carrying power of the note. Since the singer is producing the same note, it is only with the benefit of modern technology that we can appreciate just how much control we require to produce fundamental notes but with different formants. In these examples, with the note is the mid-range baritone C_3 near 130 Hz one finds, hidden in the total sound, indications of more than 30 additions to this 130 Hz note, which include the frequency doublings that define harmonics of the C_3 up to C_8 (roughly at the top of the piano). The advantages of the three tongue positions are that the presence of strong higher harmonics and partials give the note considerable carrying power (and an element of harshness), plus they demonstrate the great range of tone subtlety that is possible.

The other very obvious feature from the Figure 12.2 is not that the voice is producing harmonics of the intentionally sung note (C_3), but that there are up to 25 other frequencies being sounded that give the note the character and performance of a high-grade opera performer. As a listener we hear and recognize the fundamental and will realize that it was the C that was intended, but our brain totally fails to describe all the other detail, merely that we like the overall tone. Such complexity is also one reason why 'autotune' is not a useful feature for such sound patterns, as there is no way we could pre-programme such mixtures of partials, even for the three simple examples shown here.

The second observation is that while the fundamental C_3 seems to be firm and steady, this is not true either in intensity or frequency for many of the higher partials.

Historical Perspective of Frequency Analysis

One can appreciate the value of modern technology for voice training as it can now offer a simple tablet-based frequency analysis (i.e. a real-time Fourier transform). The example of Figure 12.2 needs very skilled interpretation, but the data are immediately useful in seeing how changes in sound production are influenced by technique. They also reveal the quality of uniform control, at a level which a student may not initially hear and understand. So yet again one can musically benefit from technological advances. Back in Figure 5.2, I displayed frequency analysed data we recorded from a mechanically bowed violin. Our violin data were taken in the early 1970s with state-of-the-art equipment and a microprocessor but nevertheless it took many minutes to record a spectrum and there was no possibility of sensing the sub-second short-term variations as seen on Figure 12.2. Nevertheless, our experiments were highly informative in showing the relative intensities of the harmonics. The same equipment further revealed unexpected features of how the harmonic content of a violin note changed depending on the string it was played on, and even the direction of the bowing action. Such features had not been appreciated previously, although in hindsight they were quite understandable because the violin structure is asymmetric so up and down bow strokes interact differently with the violin. Detailed examination of the harmonic content indicated more subtle effects caused by having the non-bowed strings being fingered on different notes. The interplay between resonances of the entire instrument are modified by these highly interactive changes.

I have also benefitted from very early analytical spectral data in figures of the harmonic content of wind instruments in the work of Harry Olson

(Musical Engineering, first published in 1952 and updated in 1967). At that time there were no computers, or instant spectral analysis systems. Olson used an electronic filter which steadily allowed different frequencies to be displayed sequentially on an oscilloscope screen, which built up the spectral pattern (in electronics this frequency filtering is called a heterodyne technique). The concept was simple and I am incredibly impressed with the wealth of data he generated by such an elegant data collection route. At that time, capturing the waveform in a single picture, and then subsequently making the Fourier analysis, would have been exceptionally difficult and the mathematics would have taken hours or days. I am mentioning this to emphasize the immense progress in electronics and computing power that can provide an on-screen real-time display as seen in Figure 12.2.

Future Developments

For me, it is clear that technology cannot play a significant role in the way we produce musical sounds in singing, except via a better understanding of voice production, and possibly better feedback during practice. YouTube and other electronic access systems will become more used as a teaching aid, and delivery of songs from home studios will increase through all the various social media sites. In popular music, the market is enormous and fashions change on the scale of a decade so the various singers and groups will emerge and fade very quickly. The classical end of the music spectrum needs many years of training, but once someone is established their career path is longer. Listening to early recordings from a century or more ago reveals many changes in fashion of what audiences favoured at different times. For the Western style music there is already a major influx of excellent singers from other parts of the world, even from areas where pentatonic or other different scales and tonalities are the norm. I therefore suspect that they will introduce new styles of singing in ways that are unfamiliar to a Western audience, but as with all exotic things this will generate a big following.

CHAPTER 13

ACOUSTICS OF CONCERT HALLS AND ROOMS

The Sound of Music

The sound of the music that we hear depends both on the place where it is made and, in the case of broadcasts or recordings, on the place where we are listening to it. This may seem a rather obvious comment until we start to consider just how important are the changes introduced by the studio or room acoustics and any electronic signal processing that has been made. We might also expect that the science of acoustics would at least mean that modern concert halls would all be excellent, not least because the subject has been studied by competent scientists and engineers for around 150 years. Unfortunately, even some of the high prestige concert halls of the last 20 years have focused on the novelty of the architecture, and the size of the seating capacity, rather than the acoustics. The design problems are remarkably difficult, and the quality of the building acoustics may be fine for some productions, but poor for others. In a large multi-seat auditorium, it then becomes incredibly difficult, or even impossible, to have a theatre which is good not only for orchestral music, but also for chamber groups, soloists and spoken conferences. Large auditoria need to be economic, and classical music may not be the highest priority (i.e. the problem may not lie with the architect and acoustic engineers).

If we look back a few hundred years, the conditions for listening to music were extremely different from now. Many instruments such as lutes, clavichords, and recorders were (and are) very quiet, and so they were adequate for playing to a small group in a room but definitely not well suited to either outdoor performances or for entertaining many people in a large building. Indeed, most of the large areas where music could be played were either in churches or the palaces of the very wealthy patrons of music. As we mentioned earlier, by the time of say Handel the private sponsorship of musical events was in decline and there was a significant shift towards performing to a paying audience. Economically this meant larger concert halls and more people. It also required more instruments and more sound, plus a genre of music that could fill and be appropriate for the larger buildings. Merely using large ecclesiastical sites was not necessarily a good option because many are built of stone, with high vaulted ceilings and consequently they have echoes, and the sound bounces around off the stone structures for a long time. Fast and detailed music is therefore not fully appreciated, unless the intent is to immerse the audience in a blurred background sea of sound. This can of course be very effective and is one of the pleasures of hearing multi-layered organ music in large echoic churches, but it is not an ideal environment for many other types of compositions. Such music will rarely transfer successfully into a home listening environment.

Construction of concert halls was therefore necessary and fashionable, and one aim was to pack in a high density of people. This posed difficulties because in terms of construction, the width of a theatre would be limited by the span of the roof timbers. The net effect was often a rectangular 'shoebox' design that was relatively narrow, and closely spaced seats. The problem of the shoebox is that the people at the far end of the box are a long way from the stage. To compensate for this, alternative geometries of more horseshoe-shaped buildings emerged, and to increase the capacity, the walls were also used for seating (e.g. with boxes or balconies). However, to maintain structural integrity the walls needed vertical supports so, rather than a continuous balcony, the structures had pillars or many individual boxes for small groups of people. The boxes have a bonus that they offered a semi-private room for the wealthy clients, and of course it meant that if the top people sat at the front of the box, they could be seen by the rest of the audience. Attendance at an opera might be as much a means of having public exposure and a fashion show as attending to hear the music. The opposite extreme was also taken in some box designs, with a mirror system so the viewer could see the stage but not be seen by the audience. There are also reports of the more enclosed boxes being used for clandestine assignations.

The building constraints on the size of the hall and seating capacity were greatly reduced in the 19th century by the advent of new structural materials such as cast iron, and then by the end of the 20th century there were steel-framed buildings and reinforced concrete, both of which allowed much greater flexibility for the architects. Cast iron construction was incorporated into massive concert halls, and a major obvious example is the Albert Hall in London which opened in 1871. The aim was to have a real statement about the scale and strength of the British Empire, and the building was designed for at least 8000 seats. It is extremely impressive in terms of size, but certainly at the time of construction was acoustically very poor, and even the opening speech by the Prince of Wales was blurred because of the echoes. Health and safety legislation has reduced the seating capacity to nearer 5500 (perhaps appropriate in view of the reduction in the scale of the Empire), and improvements in acoustics have been made. The acoustics may now be better, but basically the hall is not ideal for music, so further attempts to make acoustic improvements are still in progress.

Sound Paths from Stage to Seat

Visually, we look directly from our seat towards a performer on a stage and although light may be scattered to us from the walls and ceiling the scattered light contributes very little to the way we process and recognize the image. Ideally, we like to be fairly close to the stage because for speech and singing most people subconsciously lip read to some extent, and if the view of the lips and the sound are not well synchronized, we feel uncomfortable (even if we do not consciously know why). Similarly, we do not want to see stage actions, and then hear the music after an obvious delay. For example, between the bowing action of a violin or the beat of a conductor and the arrival of the sound. For distances under about 20 metres there is no problem, as the time from source to ear is roughly 50 milliseconds (0.05 s). Also, at that distance the intensity level is likely to be quite reasonable for most types of music. While 20 metres may seem to be quite a large distance one must remember that for a modern orchestra the width of the stage can be 20 metres, and the stage may be 10 metres deep. Such a widespread source of music causes a separate, and completely different set of acoustic problems, which I will mention later. Visual and sound disconnections are more obvious if we sit further from the action, and some people may even prefer to listen with their eyes closed.

The real acoustic difficulties, and our appreciation of the sounds, are strongly influenced by the indirect sound that reaches us after reflections.

Reflections of sound cause time delays which are subtle, as they can be either negative or positive in their effects. Reflections are in some ways quite desirable because if they reach the left and right ears at different times then we may find it improves our sense of spaciousness of the hall. Basically, we are willing to trade a slight blurring of the sound for more power and a sense of space. Sound response is totally different from the way we process visual signals. In the open air we hear the direct sound from a source, and this has most of the information, but in a room or concert hall the reflections from the walls and ceiling or even low frequencies rapidly transmitted by the floor all contribute to the signals that we try to analyse and interpret as music. Sound waves move relatively slowly through the air, so if we add extra path lengths for the reflected sounds, they can be noticeably delayed. There are several situations that we need to consider. The first is that the delay is so short that we consider this reflection to be part of the direct sound. This is good news, as we hear the extra signal just as a boost to the direct intensity and continue to recognize the direction to the performer. Longer path reflections may deliver the sound sufficiently close to the original that it is not actually blurred but instead adds character and a sense of spaciousness to the music (or speech). Alternatively, a very long path delay brings the signal to us as a distinct echo, and this is a major problem in both speech and music (e.g. as mentioned for the opening of the Albert Hall).

We therefore have a dilemma in the design of the hall as we definitely do not want any separate and clean echoes, but we do want a measurable contribution from reflected sounds that will make the music or speech persist very slightly but steadily fade away. Reflections will arrive from many directions and may involve several bounces so the intensity of this fading background should be falling smoothly. It is a key design feature, and the intensity decay is called the reverberation time. Typically, we want this persistence from say 0.8 to 2 seconds, depending on the type of sound (e.g. shorter to longer for speech, to chamber music, or symphony orchestra).

To plot the various sound paths from the stage to various seats in the hall we need to include the initial direct signal, plus the first reflections from walls and ceiling, and then the multiple reflections of the entire building. These will all have different time delays, and intensities which decay at different rates. The sketches of Figure 13.1 show how different sound paths might contribute to these components. My sketch is of a

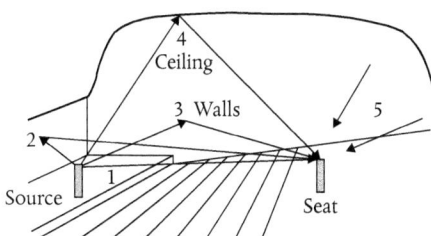

Figure 13.1 Examples of alternative direct and reflected pathways are sketched for a sound originating in the centre of the stage to a seat in the theatre. Path number 1 is direct; route 2 is reflected from the rear of the stage; 3 from the walls; 4 from the ceiling, and 5 includes background reverberation. Since the paths depend on the position on the stage and the particular seat it means that it is impossible for different seats to receive an identical signal.

very simplified auditorium, and an actual hall will include consciously added reflective surfaces, and diffusers to scatter the sound and/or kill echoes. These are reproducible but remember that the same hall acoustics are changed by the density of the audience, and even the types of clothing, so a summer audience and a winter one will influence the reverberation time, the local reflections, and the intensity.

Reverberation and Echoes

Sound continues to be heard while it bounces from different surfaces, and so modelling this reverberation at the design stage for an auditorium is essential. Modelling can be sophisticated and can include many factors of the materials and their sound reflectivity, but the design must be tested by actual performances to check that the sound is acceptable in all regions of the seating. Computer designs do not necessarily match real acoustics. An initial step of a simple measurement is to use a sound pulse to record the various arrival times, rather than playing musical notes. Ears are too subjective, and we need to collect the data with electronic systems where we can accurately look at the time dependence. Early tests used a pistol shot, but less dramatic sound pulses are now preferred. Pistol tests with an empty theatre were feasible, but not ideal, and impracticable when testing with a large audience in the auditorium (even with blanks). The sound reaching a specific seat is then recorded via a microphone, and this gives a time spread of signals from the original pulse.

Note that I am carefully saying a specific seat, as the patterns will vary between every seat in the theatre. Figure 13.2a shows the pattern that could result for a case which is musically acceptable. By this I mean the original direct signal is dominant, and so we identify the source, but any strong reflections are sufficiently close that they just add in extra power, and are not perceived as an echo. The multiple reflections of the sound around the auditorium survive for much longer, and in this figure, there is no obvious distinct echo, just a sense of persistence of the sound fading away. The important timescales involved are subjective, but broadly signals during the first 50 milliseconds are heard as initial sound. Sound delayed by more than 100 milliseconds is in the realm of a distinct echo, rather than mingling with a general reverberation. Figure 13.2b indicates the type of signal that might come from a curved ceiling. In this case there is a long additional sound path, and so the instant and ceiling signals are well separated. Not only can this cause an unwanted echo but the curved ceiling can focus the sound. Then the echo may even be stronger than the instant sound. This is a very poor result for any type of sound, music or speech.

Since we do not view or think of a hall in terms of timescales, it is better to consider distance. Here we will have some intuitive feel for potential problems, and can estimate the time delays and how much further the sound has travelled by different routes. To offer a scale of the path differences we can recall that sound travels at about 1133 feet per second, so in 100 milliseconds (i.e. 0.1sec) we are looking at distances of around 113 feet. For good acoustics, any distinct reflections must have arrived

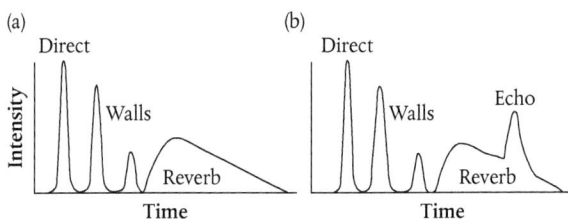

Figure 13.2 The pictures show a delay time for the direct sound, followed by discrete reflections, from walls etc., and then a general reverberation decay; (b) indicates that an echo from a high curved ceiling can be focused, but is delayed and can be strong, or even stronger than the original signal. Seating at such focal locations is very undesirable.

within about 50 milliseconds of the direct sound. It means that strong reflections must arrive via paths that may not be much longer than that of the direct sound, (i.e. no more than an extra 55 feet, or about 18 metres). Realistically, in a large auditorium many reflection paths can be far longer than this (e.g. for a mid-stalls seat, reflections from the side walls or ceiling can easily represent this extra distance). If such reflections are commonplace and feasible the hall and wall design must actively aim to reduce their intensity (reduce the sound, not kill it, or the room will sound dead).

Not all concert halls match our ideal timing pattern, and I have been to concerts in a local monastery which has a circular shaped large church with concrete walls. Architecturally it is attractive, but acoustically it can be a disaster because the echo problem is very pronounced in some areas. I heard one concert in which I was sitting near the centre of the floor area and the orchestra was backed up to one curved wall. I remember looking forward directly towards the trumpets that were playing relatively softly but the trumpet sound appeared to come very strongly from over my right shoulder. The circular curved wall not only provided a long reflection pathway, but the curvature was ideal as a lens to focus the sound to my seat, and it gave a clear delayed echo that was far stronger than the direct sound. Even though I understood the cause the effect was unsettling and it certainly impaired the music.

The same problems can occur, not only for clean reflections, but also in the overall multi-reflection patterns and general background sound. Once again it is essential that we reduce their intensity, kill any focused reflections, and intentionally scatter the sound. Rather than have zero background, we want it to fade and fall below our hearing range on the time scale of 1–2 seconds for most types of music. Needing a long but weak sound duration may seem perverse. Nevertheless, unless we have such long reverberation times, the music (or speech) will sound totally dead. Worse from the view of the architect is that the ideal reverberation times required are very different for speech and for the potentially wide range of musical styles that may be presented in the hall.

How Meaningful is Citing a Reverberation Time?

Measuring and citing a repeatable and reproducible parameter, such as reverberation time, is certainly useful as a first step. In reality, we should note that we have picked a definition of the time needed for the sound intensity to drop by a million times. This is roughly our hearing range at the more sensitive frequencies. Therefore, it is a semi-sensible choice. However, it is a far bigger span than we will hear from most speech and/or classical music. Much of the music we listen to will never be at a maximum volume initially (except for pop music), nor will all the notes be at the frequencies of our best responses. So for speech and quieter music the sound level will drop below our detection limits far faster than indicated by the cited reverberation time of the building. While this seems self-evident it is frequently not mentioned (or understood). The positive feature of this error is that normal speech may remain clear, even in a large room with a nominally long reverberation time. Since unamplified speech intensities are typically halfway up our logarithmic response scale, it means that if the cited reverberation time were to be 1.5 seconds, then the effective value for speech is only half of this (i.e. 0.75 seconds), which is a very acceptable value.

Also, for a multi-purpose auditorium, the better designs have features that adjust the decay times to suit the very varied demands of conference speeches, string quartets, operas, or symphony orchestras. All these types of sound need different reverberation time to offer sound clarity, or a good musical ambience. The ability to adjust the reverberation time is also desirable if the hall has a large seating capacity but is only partially occupied. For larger modern halls we may not realize that we rarely hear the actual sound of the performers (even for classical music) but instead one which is modified by electronics. Electronics is of course a dominant effect for pop music and jazz concerts as these have different demands, because in those cases the sound will always be via microphones and electronic amplification and speakers. The trend is that such music is not only delivered at much higher power levels, but that the weaker sounds are amplified even more than the loud ones. Overall this produces a very compressed intensity range. For the pop music delivery this is what the audience like, but for classical music (and indeed theatre) it seriously degrades the subtleties of the presentations. As already mentioned, such compression is also a standard feature of many broadcasts and CD recordings and is one of the immediately obvious sound differences between live and electronically modified performance. As listeners to continuous high intensity values, we may be less responsive to the influence of reverberation acoustics, and instead hear the music against a steady high-level background.

Similarly, outdoor area concerts are totally electronically defined in terms of intensity decay of the sounds. Even less obvious is that in such situations it is not uncommon to have pre-recorded soloists who appear to be singing but in reality all the balance and adjustments needed for outdoor sounds may have been made previously. The singers may just be miming the words, or singing to inactive microphones, while the ready-prepared music is broadcast to the fans. With long distance outdoor viewing or via local projection screens the time difference between perceived action and received sound can be kept short enough to synchronize the screen actions and loudspeaker music.

Controlling the Reverberation Time

The two key features that define any room reverberation time are the volume of the room (V), and the sound absorption (A). Reverberation time is then proportional to V/A. Hence very large rooms have very large reverberation times, and rooms with sound absorbing materials (or open windows) have shorter times. People are strong absorbers of sound, so a half-filled concert hall could potentially have almost twice the reverberation time of a full audience hall. To overcome this variability, better halls are designed with sound absorbent seating, that acoustically approximate to having a person on the seat. This reduces the differences between the hall being full, empty, or semi-full.

The packing density of the seating is also important in changing the value of A, and for example the Vienna Musikvereinsaal built in 1870 is a simple shoebox design with reflective walls and a flat ceiling. The audience is packed into this extremely densely, so this drops the effective cross section for them to act as separate sound absorbers. When the hall was designed back in 1870, the music-going Austrians were rather smaller than a modern audience. Certainly, the seat widths and leg room would not be welcomed for modern audiences from say the USA or Sweden. The very tight seating means that, although the volume is not large, the A value is less than expected for the number of seats. The overall effect is a surprisingly long reverberation time, approaching 2 seconds.

Perhaps less obvious (and again a feature that is certainly less often mentioned) is that the reverberation time of the hall also defines how long it takes a steady sound to build up to maximum power in the hall. At equilibrium sound levels, the power input from the stage matches the power that is being absorbed;, this means that for a steady intensity source the received power becomes constant. The second aspect of this reverberation-defined intensity rise is even less appreciated: the reverberation time of the stage area can be noticeably different from the main body of the hall, especially if it is designed as a smaller and semi-enclosed volume. In a good design the stage reverberation time is often ~0.1 seconds less than the main hall. For an opera set, the space may be somewhat separated from the main hall, and also be more absorbent as a result of the designs of the side features of the stage. This is desirable as it means the speed of intensity build-up at the vocal source is rapid, and it reduces the intensity variations of the initial sound that is emitted into the audience. Such a rapid feedback to the singers has a second benefit, in that it gives them a greater feeling of intimacy than they would feel from directly singing into a large auditorium.

Overall, there is a pattern which confirms that the V/A (volume/ absorption) defines the reverberation time for different types of theatre, and how it is important in the characteristic sound quality of each theatre. Sound 'quality' is mostly opinion, but a measurement of reverberation time is usually a repeatable and quantitative measurement. So there are many data taken from various halls. Table 13.1 offers a list of examples of concert halls from different eras, which shows the historic pattern of increasing seating size, and how there can be quite different reverberation times when materials, reflections, and absorbers are considered and are engineered into the overall design of the hall.

Figure 13.3 shows that typically for older buildings of similar volume the reverberation time acceptable for speech is less than around 0.8 seconds, but for opera halls of the same size, a slightly longer time is needed. Protestant churches and synagogues seem to be similar in their response, whereas concert halls of the same size have a longer resonance. Surprisingly, Catholic churches generally resonate for even longer, at maybe 1.8 seconds compared with the less than 0.8 seconds for the speech type halls of the same volume. Clearly there is potentially an interesting area for research to see if the church music written by

Table 13.1 Reverberation times and seating capacity of different concert venues

Hall	Year	Seating	Reverb. Sec
La Scala, Milan	1778	2289	1.2
Dresden	1841	1290	1.6
Opera Garnier, Paris	1861	2131	1.1
Musikvereinsaal, Vienna	1870	1670	2.0
Royal Albert Hall,	1871	8000 (5500)	5; now ~3
Kleinhans, Buffalo	1940	2839	1.5
Philharmonic, Berlin	1963	2335	2.1
Christchurch, NZ	1972	2622	2.4
Segerstrom Hall, CA	1986	2994	*adjustable
Goteborg	1994	1390	1.7
Glyndebourne	1994	1245	1.25–1.4
Kyoto Concert Hall	1995	1840	2.0
Tokyo New National	1997	1810	1.5

*Note the Segerstrom Hall has four adjustable reverberation chambers with fabrics to modify the sound absorption for different types of performance. The hall website displays an ingenious obliquely split seating layout.

composers of different faiths show differences that might result from their exposure to different church acoustics and reverberation times. Alternatively, one might ask if the same music is played at speeds that depend on the volume of the building, and if the delivery speed changes for churches depending on the denomination. A good test area might be Germany, where the language etc. is the same, but there was a north–south split in religion. My guess is that music may be played at slower speeds in the Catholic buildings. Fortunately, music can cross religious boundaries, and for example the Swiss Catholic composer Ludwig Senfl wrote motets for the Protestant Luther, so testing my speculation may be difficult.

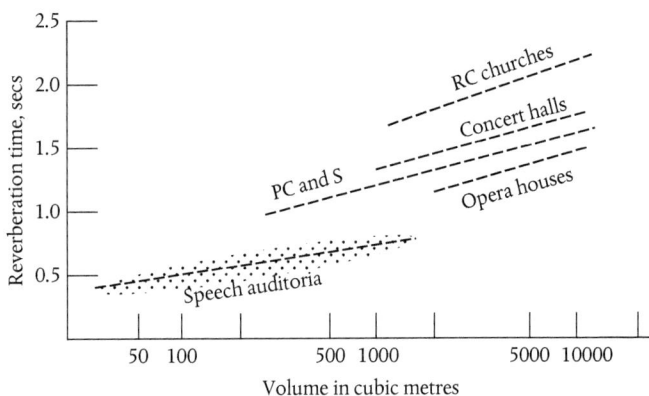

Figure 13.3 An example of data that suggest that the reverberation times of building have developed differently depending on their usage. In all cases, bigger buildings resonate for longer. Catholic religious buildings tend to have longer reverberation than the Protestant churches or Synagogues (PC and S), even though the sites may have the same interior volume. Volume is only one factor, and design and materials are critical.

Experience in Improving Acoustics

The Albert Hall was built as a status symbol, prior to much of our understanding of acoustics. It therefore offers a positive example of design and status as the centre of an Empire, but acoustically it was a disaster. The result is that it has become an acoustic laboratory in the attempts to modify the excessive echoes and make it acceptable for the various types of event that are held there. The first difficulty with a very large auditorium is loss of power at distant seats. Sound power from a source such as a metal triangle, or tuning fork, spreads equally in all directions and, rather like the light from a lamp, the surface area that it is crossing increases as the square of the distance from the source. More usefully, it means the sound intensity drops as the square of the distance. The loss of sound intensity with distance in the large concert halls, such as the Albert Hall or the Metropolitan Opera House is a real challenge. In the latter there are seats some 60 metres (~200 feet) from the stage (i.e. two-thirds of a football pitch). For viewing of the performers, stage craft, and actions, as well as for hearing the sounds, such enormous buildings are far from ideal. If we compare the sound energy for someone sitting in the front of the stalls, at 25 feet from the triangle player in the orchestra, the sound is 16 times more powerful than for someone at the back of a modest size auditorium, who is 100 feet from the triangle (i.e. 100/25 = 4; and 4^2 =16). This power loss falls to a mere 1/64 of the original power for the 200 feet distant seat. This loss of information is probably far greater than the difference in seat prices at these two extremes.

At such long distances, many of the audience may use opera glasses. This will certainly help with the view, but there will be a discernible time lag between the singers' lips, which we see via the opera glasses, and actually hearing the sound some 170 milliseconds later. To emphasize the scale of this delay, a fast Shakespearean actor may speak dialogue at several words per second, so the sight to sound gap can be equivalent to about one entire word (or more).

Concert halls not only deliver direct sound but also reflect power, and here the Albert Hall is particularly interesting. The layout of the hall is sketched in Figure 13.4, which indicates the stage, an arena, plus seating in banks, and box structures around the edge. The building interior is effectively an ellipse with the two foci on the plane of the stage, and on

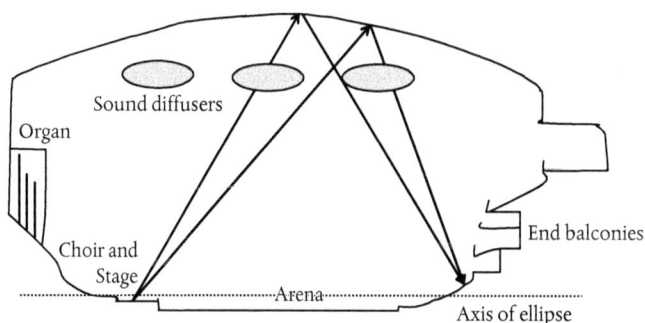

Figure 13.4 The sketch of the interior of the Albert Hall emphasizes that the ceiling is part of an ellipse with the foci and main axis close to the level of the stage. The extreme size means that there are very long paths and delays in sound arrival from the stage to the rear of the theatre. Because of the elliptical shape, the roof-reflected signals are not merely delayed, but are focused to be very strong. Indeed, they may be far stronger than the original direct signal. The shaded ovals indicate that there have been additions of items to scatter and diffuse the sounds. Level changes and balcony sections are also relevant, as they add reflections.

the rear seating. This causes a very strong and localized imaging of the sound between these two regions, but because the entire roof structure acts as a lens to focus sound from one focal point to the other, the intensity via the original roof reflections was far stronger, but delayed, compared with the direct sound. Similar effects from the curved ceiling apply to other seating areas.

The transit time for a direct signal along the length of the hall can be around 100 milliseconds (for ~36 metres), but a ceiling reflection route can take more than 200 milliseconds (i.e. a very long delayed but intense echo). One of the many options to reduce this problem was the addition of structures hung from the ceiling (variously called clouds or inverted mushrooms) that scatter the sound to reduce the focus effect, but that allow some sound energy to go to the ceiling to contribute to the background reverberation. Overall, this has reduced the clear echo, and energy is dissipated to drop the reverberation time down below 3 seconds, but this is still too long for some types of music. The long ranges involved weaken the sound quality, both in the arena and in the more expensive seating. All areas suffer from the distinctive acoustics of this large building.

Broadcast performances from the hall will disguise many of the problems, as with careful microphone positioning, the separate sources do not have time lags, and the directionality of the microphones can minimize the long reverberations. The performers will normally still feel that they are playing into a vast void. This can be very obvious, even on the televised versions because one senses that soloists are over-powerful in their attempts to increase their volume, and for many singers the soloists perform below their best because of the unfamiliar acoustics.

There is a saying that if a composer has a new work performed in the Albert Hall, then a second, or third, hearing is automatically guaranteed.

In the next chapter I will consider the scale of time delays caused by the use of large orchestras, but the Royal Albert Hall offers a convenient introduction to the problem. The stage dimensions start at a 'narrow' 13 metres, but by the seating of the chorus it has expanded to around 25 metres wide, and from the conductor to the organist there is a gap of around 13 metres. With spacings on this scale it is inevitable that for people seated on one side of the auditorium (or choir), the various notes (hopefully all produced initially at the same time) will have measurable and noticeable time delays merely from the vast spread of the orchestra, choir, and organist. In fast passages the distance separations between them blur the notes, even without the very large delays from the building reflections. For those who like numerical examples, we can estimate that for a metronome speed of 180 crotchets per minute (fairly fast), and notes written as semiquavers, then merely crossing the orchestral stage means there will be overlap of the notes from one side to the other.

Not all Victorian halls were built on the same massive scale, and many were of high quality for music and opera or vaudeville. Excellent buildings were designed by Frank Matcham and his colleagues. He was a pioneer in the use of steel girders to allow the construction of cantilevered balconies, which avoided the earlier problems of intrusive pillars. His sense of acoustic design was good, and his fire and safety features meant that many of his theatres survived for a very long time.

By contrast there was a 19th/20th century pattern of designing seaside theatres with a flat floor and a rectangular floor plan. The flat floor increased the volume, and so lengthened the reverberation in the building, but this was accepted for the ability to use it as a dance floor. Presumably, for the same reason, the stage was often set on the long side of the rectangle. Musically this is quite a poor design voices and many of the instruments would only be clearly directed at a small fraction of the audience.

How to Mix Soloists and Orchestra for an Opera

The layout of the stage area is a particular difficulty with opera. The problem is twofold. The first is that, because not only the sound but also the stage actions are important it is desirable to have the singers as far forward towards the audience as possible, to improve the intimacy of the action in the opera. This is in conflict with a design that has the orchestra in front of the stage, at a level and sighting position that gives good orchestral acoustics to the audience. This is true even though the opera orchestra may be quite small compared with a large symphony orchestra. The typical solution is to have the orchestra in a pit below the level of the stage. Acoustically, for their contributions to the music this is not ideal. If it is set at a lower level in front of the stage it distances the stage from the audience. A solution adopted by Wagner in the design of the Bayreuth opera house was to have not just a deep pit, but one which ran back underneath the stage (Figure 13.5). His intent was that the orchestral

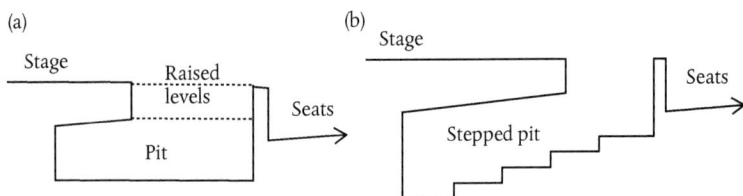

Figure 13.5 There are several patterns for accommodating an orchestra with an opera performance. These include a seating where the audience can see the musicians, but this separates the singers from the audience. Alternatively, the instruments can be in a pit beneath the stage. This advances the singers, but the orchestral sound must reflect onto the stage before reaching the audience. (a) An orchestra slightly under the stage, but with options of raising part of the pit floor to higher levels when smaller groups are used. (b) A design with a very deep and large stepped orchestral pit.

sound would be reflected back to the singers, and only escape to the audience after reflections and scattering. This pit geometry may be good for the singers and acting but musically it seriously degrades the orchestral sound quality and destroys any audience sight line of the players. It also means that for many seats in the auditorium all the orchestral sound is heard via reflections. This introduces small but definitely noticeable time lags between the singers and the musicians. It is especially obvious if one can see the conductor as the baton action clearly precedes the sound of the music. In such opera houses there is a loss of sensitivity to the playing of the musicians, and the layout weakens the sound quality of the pit orchestra. Opera may be marketed in terms of the solo singers, but loss of orchestral background quality undermines the soloists as well. The pit system is often used for ballet, and here it definitely reduces the impact of the orchestral music, which in this case is crucial.

An improvement in pit design exists in some opera houses, such as Glyndebourne in England, where the base of the pit includes a moveable platform. It can be at full depth and can accommodate an extended orchestra stretching beneath the stage. This is acoustically poor, but allows the orchestra to have a wide range of instruments. Raising the platform to a partial depth is feasible for smaller orchestral numbers, and this benefits the sound projection. Alternatively, the base can be raised fully to stage level when the hall is used for other types of performance.

An additional comment on opera house designs that incorporate vertical side walls lined with enclosed boxes is that the designs frequently deaden the reflections from the side. This may be useful in reducing the reverberation time but it interferes with the ambience and spaciousness of the sound in the main auditorium. Within the boxes the view of the stage can be good but certainly at the rear of the boxes the acoustics are not ideal. This even applies to such famous houses as La Scala in Milan.

Concert Hall Reflectors, Clouds, and Sound Diffusers

Instead of the original rectangular (shoebox) seating area, there has steadily been expansion in seating areas backwards, sideways, and upwards. Angling the floor is visually useful as it offers better sight lines for the more distant seats, and acoustically it is good as it reduces the total volume of the hall and minimizes the reverberation time of very large auditoria. Floor plans such as the expanding fan pattern offer more seating with a view of the stage and relatively good proximity. The fan angle is always a point of conflict, as a large angle gives more seating but acoustically it has limitations, particularly as singers project their voices forward with a rapid angular fall off in sideways-going sound. Further, wide fan shape seating can cause strong but delayed reflections. Such conflicts define the limits of the angle of the fan. Side boxes, as found in earlier opera houses around horseshoe floor plans, pack in a larger audience, but the boxes may have poor sightlines, even if the acoustics are adequate. Covering the sidewalls with boxes can seriously reduce the essential reflections into the main auditorium.

Additions of balconies offer more seating, but delivering sound to seats under balconies may require deliberate reflections so the wall and reflector designs are important. These need some careful thought because under-balcony seats often receive a poor quality and filtered sound. Ideally the reflectors from the stage should help to send sound directly into such under-balcony areas. Even better is to angle the underside of the balcony to offer a larger entrance area and allow a downward reflection within that space. There are architectural guidelines that suggest how deep an under-balcony seating can exist relative to the throat and entrance shape beneath the balcony. These comments are typical of the considerations that have been made in concert hall design where acoustic properties were considered and/or attempts made to add useful

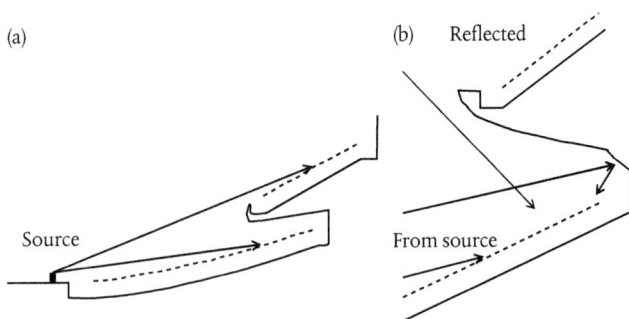

Figure 13.6 Examples of direct and reflected sound paths for good balcony designs. Seating is indicated by the dots. In (a) direct sound can reach all the seats, whereas in (b) reflections are needed for the rear under-balcony seats. In both cases, the reverberation times for under the balcony seating are impaired.

reflections. Some of the schemes and designs are briefly sketched in Figure 13.6. Note that the care of balcony design has often been ignored in theatres designed for films, as there the screen view is paramount, and all the sound is delivered by speakers, which can also be placed beneath the balcony level. Such auditoria are therefore adequate for cinema but can be quite poor in the under-balcony areas for speech or concerts.

Modern architectural ingenuity in packing large audiences into moderately close proximity to a stage are extremely varied (both in layout and acoustic success) but numerous websites offer examples to show and discuss the relevant floor plans and images (e.g. as mentioned for the unusual 'Segerstrom Center for the Arts' design).

Listening to Music at Home or in a Car

Home listening comes in two varieties: where the music is being played or sung in the room and where the sound originates via some electronic system (e.g. radio, TV, or CD). The two situations are very different even for the same piece of music (e.g. singer, pianist, quartet, jazz, or pop). The home room reverberation times are highly subjective and depend not just on the room size but critically on the absorption and reflection properties of the carpets, curtains, furniture, and walls. The obvious demonstration example is to compare singing in a living room, and singing in a bathroom of a similar size. The echoic bathroom will produce a vastly different performance both in decay time and in the relative intensities of the notes at different pitch. Lower male voices may resonate better but higher soprano notes will sound excessively strong and bright. In part, this may be that for low notes the male chest cavity is too small and the echoic bathroom will compensate for this. The differences between rooms (and concert halls) are inevitable and appear no matter who is performing.

There are some bonus points for playing in a house size room. The first is that we have a feeling of immediacy and intimacy with the music, the sound level is of course higher as the distances are small, and there are no visual delays between seeing the action and hearing it. We also hear the full frequency response of the notes as they are produced. The downside comes via the reverberation, which can be very dead, and the damping of different frequencies will depend on the properties of the materials that cause the reflections and absorption. None of these effects are unexpected but they are surprisingly important for the enjoyment of the music. For example, the rooms in my house are not unusual in size but some have carpets and soft furnishing, while others do not. If I sing or play a violin in the different areas the apparent quality of the performances (and enjoyment) is noticeably different, both for me and for anyone who is forced to listen. No room offers the sound I would expect in a concert hall. Experiments with portable equipment show that the sound of my radio or CDs differs between rooms far more than I would intuitively have guessed. There are also differences between reproduction on alternative electronics and speakers (as expected). I am surprised that I definitely prefer some CD music in different rooms on different vintage equipment. Such differences emphasize for me the divergence in recording techniques during the manufacture of the CD. They also demonstrate how sensitive and critical we are to changes in the acoustics of both the electronics and of the reverberation and ambience of the room.

By contrast, listening to music in a car or with other high-level background means we can never appreciate music with a wide dynamic intensity range and long reverberation times. In these situations the radio stations that compress the intensity only from fairly loud to very loud are acceptable. It definitely is not the music that was performed but it is a delivery that is suitable for the acoustic conditions where we are listening. In the UK, there are very clear examples of relatively original sound dynamics (BBC 3), and music that is strongly compressed (Classic FM). I am sure that I am typical in that I prefer the compressed sound when driving, but the uncompressed version, when listening quietly at home.

The only conclusions that we can draw from the preceding comments are that no two concert halls give the same musical sound; no two seats in any concert hall give identical quality music; and there is absolutely no way that live music will sound the same in a home environment compared with a concert hall (even for quartets and soloists). Broadcast and recorded music played at home will depend not only on the equipment but also the room, and almost always the result is a poorer version of the signal being offered. Many people accept this and either try to remove the room effects with high grade headphones or invest much more money on a dedicated listening room with more exotic electronics. Finally, no matter how we describe the differences to other people they will never fully understand. All of us have unique hearing and musical backgrounds. For your tastes you are the expert.

CHAPTER 14

ORCHESTRAL LAYOUT AND
THE BEST CONCERT SEATS

Orchestral Size

O rchestral size expanded considerably from the small orchestras of 12 to 16 of Bach or Haydn, to the much larger group of 40 people available for Beethoven by the time of his last symphony. By contrast, the big modern orchestras will often routinely include a hundred performers, and special events have yet more players; additions of major choirs can add a further one or two hundred singers. A modern orchestra may therefore have more first violins than an entire ensemble at the time of Bach. The larger numbers occupy a far greater stage area and, as was obvious from the preceding discussions, there are inevitable problems of time delays, echoes, and different reflections from the various parts of the orchestra to the members of the audience. For a critical member of the audience the sound will noticeably depend both on the seat that has been chosen, and the layout of the orchestra. This can be confusing, as although it is easy to look at the seating plan (and the prices), there is no guarantee that each orchestra will use an identical layout of the instruments.

Orchestral size will equally influence both the conductor and the feedback between instruments for the members of the orchestra. High speed rendition may be the aim of an extrovert soloist, but the performance will lose clarity if the response time of the large orchestra (and concert hall) cannot match this tempo. It is also necessary to recognize that the conductor and soloist are at the centre of the stage, and so they are in the ideal position to hear all the instruments at a minimum distance. Further, they are relatively isolated from many of the acoustic feedback features of the concert hall, which means that any long-range reflections will have dropped in volume or be blurred into a background. The conductor/soloist positioning can therefore easily lead to music that is too fast for the audience, both from the time delays across the orchestra and for the response of the auditorium. Having played in various orchestras and groups, I have sensed that there are quite large differences in the way I respond to the others, depending on the seating. Front desk of the violins is fine, but I have definitely felt less involved and distanced from the action in other locations. This feeling may even exist for professional players.

A good conductor is likely to occasionally venture out into the auditorium during rehearsal to see how the sound is delivered to the audience, rather than to the podium. Not all conductors recognize this, and I have attended many concerts, particularly with large choirs, where the speed was excessive for the building, and the result was just a blurred wall of sound to the audience, but perhaps less so for the choir or orchestral members.

I suspect the problem is compounded by recorded versions of a performance, as these will use microphone arrays that offer instant inputs from the various orchestral sections and these minimize the time delays of the signals to the recording tape. Effectively they reduce the time spread caused by a large orchestra (i.e. they offer the podium view of the music). Hence the soloist and conductor may be far happier with their recording than the live audience was with the performance.

Time Spreads Across a Big Orchestra

Rather than repeat that modern orchestras and choirs can be large, we can be quantitative about the time lag. I previously calculated that for a big orchestra the time spread across it would often correspond to one note, or more, in a fast passage. This is equally true for the audience out in the auditorium. Differences from various parts of the orchestra and choir may be as great as 0.1 seconds (100 milliseconds). This may not sound like a big number, but in sports such times make the difference between winning and losing. A delay of 0.1 seconds is easily discernible in a top 100 metre sprint, as it means a difference of one metre, or several places in the race. In fencing epee, the difference between one person scoring, or both registering during a double action is 25 milliseconds. This short time interval is easily recognizable by the fencers. In other words, these differences of fractions of a second are noticeable, both physically and acoustically. For fast musical passages time delays cause a permanent dissonance for a listener seated on the side of the concert hall (not forgetting that the problem also exists for the performers).

Time delays caused by reflections introduce slightly different problems, as the brain initially processes the direction and content of direct sound and treats reflections somewhat differently. This seems a reasonable evolutionary strategy for survival as we (and other animals) need to worry about immediate threats and dangers as a top priority, whereas as delayed information from reflections is already being processed by the brain.

Identification by Direct Sound

As part of our evolutionary need to survive, and avoid being eaten by a lion, we not only had to recognize the directions of the sound source, but also had to do so sufficiently quickly that we could react and identify what (or who) was producing the noise. In all cases, whether speech of individuals or musical instruments, we start to do this acoustically well within the first 50 milliseconds, by listening to the way the sound intensity builds up, and simultaneously by doing an analysis of the component frequencies of the sounds. This brain processing is incredibly impressive, and modern high-speed electronics and computing have only just approached this type of high quality response on such short timescales, and certainly still cannot do it with such a small 30 watt multitasking processor (the brain).

Our first step is to see how the overall intensity changes with time. Reasons for the differences between different instruments are intuitively obvious. A struck triangle will leap to maximum intensity and fade quickly, whereas an organ note will rise quite slowly up to the peak power as it takes time to activate all the air in the organ pipe. By contrast a piano note intensity will rise quickly when the soft hammer hits the strings, but the sound then fades rather slowly as energy resonates between the strings of the piano. The identification of an instrumental sound type is totally subconscious. In the first few critical tens of milliseconds our brain detects and analyses this build-up of power, and the memory section identifies what instrument is being played. The form of this power characteristic is called 'the starting transient'. Two examples are sketched in Figure 14.1, for a piano and an organ. I have picked this pair as they have very different time-dependence because of the mechanics of the sound generation.

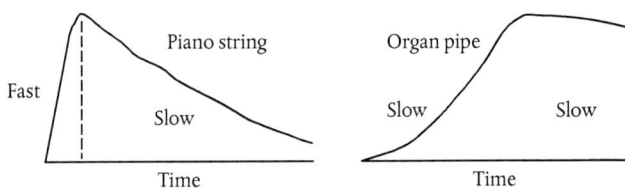

Figure 14.1 The manner in which the power starts up from a piano string and an organ pipe are very different, and these starting transients identify the instrument for us. The piano hammer blow gives a rapid rise in intensity, but then an undamped note will resonate for tens of seconds. The power does not fade smoothly, but has some oscillations in intensity, particularly when there is more than one string per note. In complete contrast, it takes time to activate the air in an organ pipe, so the power rise is slow, but it can be sustained, and then falls quickly when the air flow is stopped. The time patterns are effectively opposite extremes.

Our instincts and identification of the instrument can be confused if we deliberately alter the form of the starting transient. In music courses, I have used a standard trick to show how important this is. I have asked the students to identify the composer of a piece of music, and the instrument that was used. In my primitive example, I used a tape recording of a slow piece of Bach played on the piano. However, instead of playing the tape forward, I reversed it. It started at the end and played towards the beginning. Incredibly most groups have correctly identified it as music by Bach, but virtually without exception, they said it was some type of early organ. A glance at Figure 14.1 shows the reason why. For the notes on the piano the intensity rises rapidly and decays quite slowly, whereas an organ note rises very slowly. Hence, a backward piano note resembles the slow starting transient of an organ.

I must admit that I generally said 'organize your thoughts and identify the instrument and the composer', subconscious priming of *organize* was intended to influence the result.

Analysis of the Spectrum of Component Notes

In earlier chapters I mentioned that every instrument has a different formant (i.e. a mixture of different frequencies), when a particular note is played. The thin violin string was particularly simple as the vibration of the string in a sawtooth type motion could only occur if the mixture had a series of harmonics of 1, 2, 3, 4 etc. times the fundamental frequency. From the isolated string these fell in power as $1:1/4:1/9:1/25$ etc. Because of the frequency-dependent amplifier characteristics of the violin, the sound we hear had all the harmonics, but their relative intensities were altered compared with the unamplified string. Nevertheless, when we hear such a sound mixture, we recognize it as belonging to one of the violin family. Control of these responses separates the 'good' from the 'average' quality instrument.

Formant processing and memory allow us to distinguish all the instruments of the orchestra, as well as the voices of people we speak to. Voices of the same family are often very similar, in the same way that English (or Spanish) differ somewhat between the various countries that use the languages. In general, we can normally rapidly distinguish between individuals, or identify someone's country of origin. Our voiceprints are sufficiently well defined that they are used in forensic identifications. We find this identification much easier if we have a full frequency range. Listening to a voice on a telephone, where there is a relatively limited frequency response, can cause problems in some cases. If my daughters phone me and say 'It's me', I sometimes need a few words to be sure which of them is speaking.

Similar problems exist for elephants, as they can communicate over very long distances of more than a kilometre but only the low frequencies travel well. So apparently at close quarters they are able to identify individuals but at long range they are also unsure who is 'speaking'. In earlier telephone parlance would these have been trunk calls?

Overall, our extreme sensitivity and ability to distinguish individuals and specific instruments implies that we do not want the acoustics of a building, broadcast or recording to alter the original sound. We actively want to hear the variations between sounds from different directions or performers.

Factors that Change the Sound of Instruments

Musical instruments are works of art, and therefore, like people, have individuality in sound and/or ease of playing. For a violinist, the same music played on a Stradivarius or a modern violin will always be different in some parts of the range. Other differences, such as the responsiveness and ease of sound production, may only be apparent to the violinist. Nevertheless, we will all recognize that a violin is being played. What is less obvious is that if we hear the same music from a different direction relative to the instrument, then we will not doubt that it was the same piece, but in reality there will always be a different tonal balance in terms of intensity and content of the partials.

Virtually all instruments produce patterns of sound which not only differ with direction but have a different radiation pattern for every note. The exceptions include percussion items, such as a triangle. Hearing an instrument head on (e.g. directly looking at the horn of a trumpet) will never be the same as hearing it from the side or from behind. In some cases, this is familiar territory. For seats behind the orchestra (for example, in the choir seats) with a grand piano at the front of the stage, the low piano notes will boom forth in all directions but the higher notes will be reflected forward towards the main audience by the raised lid, but screened from the people in the choir stalls. This is still true, even from the side, where the seats may offer a view of the keyboard. The piano lid reflector effect is a clear example of note-dependent directionality, which exists for all instruments (and is extremely obvious for singers).

Voice projection is very much in the forward direction for higher notes, and indeed I have mentioned that this is one of the reasons why auditoria tend only to be fan shaped over a modest angle, to allow singing to reach all the audience. Broadly, we expect instruments to have directionality if they have an obvious feature which defines which way the sound energy is going. Hence, we can predict that all the brass and wind instruments, which have horns at their ends, are going to radiate much of their energy in a forward direction from the horn. The general pattern, as with people, is that the higher frequency notes are the most directional, and this includes the higher partials of lower notes. The consequence is that if one is facing a clarinet there is a strong forward sound over the whole range, but from the sides one not only hears high notes that are much weaker, but also there is a reduction in all the higher partials of every note that is being played. Effectively the head on and side on regions receive music as though different instruments were being played. This is very obvious from a military band, because in the open air we have minimal additional reflections that might strengthen high frequencies to the side. A marching band tone character will change quite distinctly as it approaches and passes us.

Instruments such as a French horn or a tuba are interesting because the horn is never directed towards the audience. Instead, it is facing sideways for the horn and upwards for the tuba. In both cases the low notes broadcast fairly uniformly (so it does not matter) but higher notes arrive after reflection from the walls (for the horn) or from the ceiling (for the tuba).

While we could make reasonable predictions of the emission pattern for the wind instruments, there is a much bigger challenge to guess at the radiation patterns from the string instruments. For some notes the air is excited by moving in and out of the sound holes (f holes). If the air movement in the 'box' is the main resonator (e.g. near the D string for a violin) then the movement of pressure waves through the sound holes defines the broadcast directions of those notes. By contrast, the vibrations of the belly and back of the instruments have more drum like vibrational patterns, and these notes will send sounds in different directions. The changes in pattern with note has an unexpected consequence. It means that even the player does not hear the same set of harmonics and their intensities as people in the audience. Further they all hear slightly different music, depending on where they are seated. The differences are real and measurable, but to various extents the patterns will be enhanced or reduced by reflections within the hall. Even listening at home, in a normal room, one can hear the alterations in tone quality caused by moving around relative to a string instrument.

Any radiation pattern that changes with frequency causes a complicated situation. Not only does a listener who faces the instrument (say a cello) hear shifts in intensity response with frequency, but also, in reality, every note produces harmonics, and their relative values will also vary in intensity with direction. Similarly, for chords the balance of component notes will differ with listening direction. It is worth realizing that the performer hears a different, and unique, sound from the instrument compared with that heard by *any* member of the audience. This result can be quite disconcerting for a soloist who subsequently listens to a recording of the performance. The positive side of such features is that every orchestral member is contributing a different and individual sound, and so we hear the composite chorus of their instrumental voices, which we appreciate, and it may associate either to a particular player, or the sound we associate with a specific orchestra.

One does not need to play an instrument to sense the difference between personal and recorded sound, because it is equally obvious if we listen to a recording of our own voice. For most of us it is strange, compared with the voice we hear during speaking.

Directionality of Instruments

We can record and quantify these directional effects by measuring and comparing instruments playing a different set of notes. As an example, I want to contrast the emission patterns from a trumpet, a French horn, and a tuba. The trumpet has a bell directed towards the section of the audience that the trumpeter is facing. The sound projects forward towards these chosen few. For people to one side, the power of the trumpet falls quite noticeably, even if they are at the same distance from the player. In each case the sound travelling in the backwards direction will be blocked by the player, and the sideways volume may be reduced by the arms. Many soloists on say that clarinets or trumpets will change the direction of the instrument when they are highlighting some phrases or technical virtuosity. Rarely will they realize that aiming the instrument skywards will actually diminish the power reaching the audience and take away the brightness of the sound.

The French horn has a bell which is directed off to one side, so there is an immediate physical reason why the radiation pattern out towards the audience is skewed to one side. It clearly makes a big difference if the French horn is positioned on the left or right of the stage. Sound from a tuba behaves rather similarly, and it will be sensitive to the positioning of the ceiling reflectors and sound diffusers.

As a simplistic guide, Figure 14.2 show some approximate sketches of radiation patterns at several frequencies. At the lowest notes, ~200 Hz (G_3), all three instruments radiate relatively uniformly in the forward direction. By contrast, all three become progressively more directional at their higher frequencies. On the figure the circles are spaced as 10 dB contour lines. Since a 10 dB volume change is interpreted as a doubling in perceived loudness the three lines span a range of at least four times in the way we hear the volume. In reality it is a drop of 100 times in true power. Note that the figure is highly schematic and will depend on both the instrument and the volume at which it is being played. Reference back to Figure 4.5 also shows that our sensitivity is very different across the sound spectrum, and it changes with age, so not only will the power

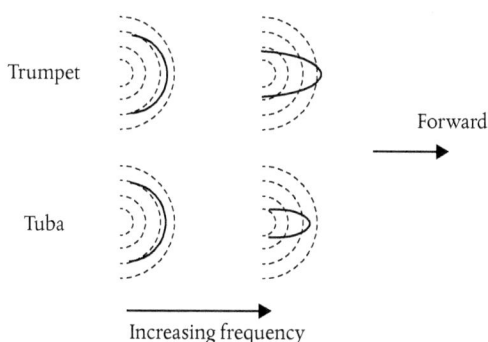

Figure 14.2 These sketches indicate changes in the forward directional intensity patterns for brass instruments with frequency. Both the large tuba and the smaller trumpet emit more intensely in the forward direction at higher frequencies. Perhaps surprisingly, the tuba becomes more focused than the trumpet at the extreme top of its range. The dashed lines indicate relative doublings of perceived intensity. Broadly, this sketch is typical of all brass instruments.

level alter, but so will the tonal balance. It can be significantly different as a result of changes in the perceived formant with direction.

The wind instrument examples resemble sound production by the voice, in that there is a clearly defined forward power output. By contrast, the vibrating plates of the string instruments show a range of directional properties, depending strongly on the frequency. For each instrument the lower register notes are also mostly radiated uniformly, but at the higher frequencies they have power distributions with angle that have strong lobes in some directions. (N.B. Their mechanical structures are asymmetric.) This is consistent with the way the plates (particularly the front plate of a cello) vibrate, and how there is interference from different parts, and so the waves can add or cancel to give a heart-shaped power pattern with angle. An example for a cello is sketched in Figure 14.3, for several notes. For the lower notes the sound is broadcast roughly in all directions around the player, but moving to higher notes means that sound can variously be directed at say 45 degrees to each side of the belly of the instrument, or only in a forward cone of power. There are equivalent type variations in power patterns in the vertical direction which will influence reflections from the floor and ceiling structures. The radiation patterns of all instruments have these features, and it is impossible to pick a single direction which will give the peak output from every note,

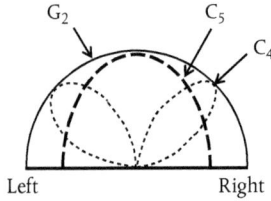

Figure 14.3 The figure sketches three examples of the radiation type patterns for a cello. At the lowest frequency of ~100 Hz (G_2), power is sent uniformly in all directions around the cello. At C_4 (~261 Hz) there are two power lobes, which may not be of equal intensity. Just one octave higher, by C_5 (~523 Hz) most power is again directed forward, with minimal power heading off to the sides or behind the cello. Left and right refer to the left and right side as viewed by the cellist. Note the cellist's bowing arm may block some of the right-hand-side signals.

and the patterns change for all the associated harmonics. This is not only true, and a problem for the audience, but is an equally challenging problem for a recording engineer, even when using many microphones.

Arrangements of the Orchestra

Since the projection of the sound is so variable with note and instrument these are factors that will influence the sound of an orchestra as it is heard in the auditorium. We therefore need to look at typical orchestral layouts, and try to assess if they have predictable and measurable effects on the sound. The fact that different orchestral arrangements are in use suggests that at least for the conductor, or sound engineer, these are important factors, but there is not necessarily agreement as to which is ideal. Also, different layouts may be desirable for different types of music, for specific concert halls, or in recording studios. Figure 14.4 sketches two of the more common seating plans for an orchestra.

The sound radiation patterns pose problems not only for the audience but also for deciding on the best seating placements of the instruments within the orchestra. While first violins tend always to be on the left side of the stage, there are many variants with placing of the rest of the string instruments. Figure 14.4 indicates significant differences between the more common seating plans. These various permutations alter the sound heard by the conductor in terms of balance of instruments and create very noticeable changes to the sound heard by the audience in the

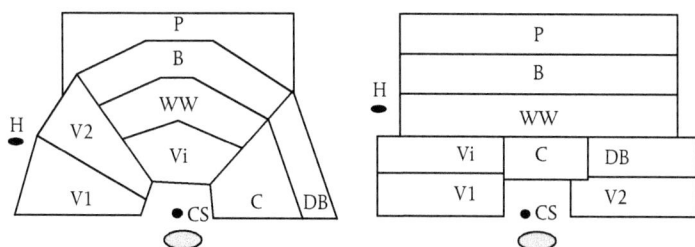

Figure 14.4 This figure contrasts two patterns of the seating layout for the main orchestral instruments for commonly used stage arrangements for public performance. Many other layouts may be used in recording sessions, or as determined by the design of the stage or by conductor's preference. Some plans have seats set in arcs around the conductor; others are in blocks parallel to the front of the stage. This pair contrasts the layout of violins all on one side, or with firsts and seconds facing one another. Labels are: CS conductor and/or soloists; V1 and V2 first and second violins; Vi violas; C cellos; DB double basses; WW woodwind; B brass; P percussion, and H harp. Note that some stages are flat, while others have stepped sections, and these features influence both the layout and the sound.

concert hall or signals via microphones. Violins are a key feature of most symphonic music, and they are relatively quiet instruments. Hence there may be a dozen of each in both the first and second sections. They are essential, but because the string instruments are of low power, they must be near the front of the orchestral platform. Two common seating patterns are for all the violins to be on the left side, and the cellos and double basses on the right, or for the first and second violins to face one another with the cellos more centrally placed. The pattern with violins facing one another is sometimes called a German plan, as it was one used by Wagner. The cellos-to-the-right arrangement has been termed an American seating, but even within these patterns, variants are used.

If all the violins are to the left side, then first and seconds have a similar sound. The emission patterns are not symmetric about the violin. Typically, it favours the bowing side for notes near the frequency of the open E string (660 Hz), but this shifts to power to the left for notes an octave or so higher (1320 Hz). The right-hand arrangement of Figure 14.4, with the second violins to the right of the stage, produces an orchestra with a second violin sound that is stronger at high frequencies compared with first violins. The sounds will still be obviously from violins on each side, but with a different tonal quality. One could argue that this is desirable, as the two violin parts will be more readily identifiable (i.e. the placing gives them different tone quality). One can then claim that Wagner's 'German' seating plan has different violin tone quality and distinguishes between the firsts and seconds. Such a division is a bonus. Having now understood the frequency-dependent radiation patterns (something unknown to Wagner), it is obvious that there is a serious flaw in this arrangement. Instead, we could alternatively propose a totally heretical argument that, because the first violins tend to play higher notes compared with the seconds, and high violin notes are projected to the *left* of the instrument, then the *first* violins should be on the right side! I suspect one reason against this ever being considered is that the view of the first violinists would be impaired, and just as for a soloist, we in the audience want to see the faces and bowing action of the players. The main objection is that it is not the traditional layout. However, in a recording studio it might be a valuable experiment.

Reseating the cellos will similarly alter the cello tone and as sketched in Figure 14.3, there will be major differences in relative power of notes at say C_4 and C_5, even for the audience in the central stalls. If we assume that the central stalls are among the higher priced seats, and/or attract the more musically aware, then this is a serious consideration. Equivalent changes emerge for all the other instruments. The only easy conclusion is that no layout pattern is perfect, and there are fundamental reasons why perfection can never exist.

The type of stage can alter the balance between the instrumental sounds and there is a clear difference between using a flat stage and one that is stepped. In part, the tone is changed because many wind instruments are angled downwards when played, and also the sound of rear instruments will be blocked by those in front on a flat stage. As mentioned earlier, the brass instruments (e.g. trumpets and trombones) can produce a highly directional beam of sound which in fortissimo passages is well above the legal safe intensity limit for the hearing of the musicians seated immediately in front of them. As discussed in Chapter 4, exposure to such sound levels produces hearing loss in a relatively short time. This has now been recognized, and in some countries the seating plan, and instrument spacing, has to fulfil health and safety standards to avoid sound overload, and the deafness created in the unfortunate musicians sitting directly in line with the brass. Recent successful compensation claims will guarantee that changes will appear. Legislation will probably do nothing for the damage inflicted at high volume pop concerts and discos. Even young people who regularly attend discos tend to have impaired hearing levels, that previously would only have been associated with old age. This is sad as it applies to millions of people.

In all types of music, soloists tend to be placed at the centre front of the stage. There are inevitably exceptions to this rule and I have seen the four choral soloists of the Beethoven 9th symphony placed behind the orchestra, and in front of the choir. I assume this was a placing forced by the relatively small scale of the Viennese stage, compared with the size of the choir and orchestra, but the overall result was very unfortunate. The soloists sounded extremely remote and were overshadowed both by the choir and the orchestra. In general, in live performances (without electronic amplification) most solo instruments or singers can struggle to be dominant when competing with the size of modern orchestras. This type of power imbalance between soloists and orchestra *should* have a real feedback effect on the composer. When the soloist lacks the power to easily be heard, many composers have aimed at a dialogue, where in solo passages the orchestration is very light, or with mutes on the violins. As a violinist, I have listened to many violin concertos, and realize that such a dialogue approach is common in the performances of many popular violin works. However, when the composer has failed to recognize the problem, and generated an unsustainable power competition between soloist and orchestra, the live performances are very unsatisfying (and the concertos are deemed to be less popular). This does not mean that there are not fashionable recordings and broadcasts, as once electronics is introduced into the music, the soloist can be amplified to a level that is impossible from the original instrument. Overall this means that some violin concertos are worth hearing in the live version, but others are only enjoyable via a CD. Problems of dynamic power balance are less serious for a concert grand piano, as this is a powerful instrument, but the application of the energy needed in the competition, can often come at the expense of a less sensitive interpretation.

I am deeply concerned that there is a growing idea that soloists should be amplified electronically, not just on a CD, but even in the auditorium. In many large concert halls, it is already standard practice to have some electronic amplification. It is hidden under the guise of controlling overall sound levels and reverberation. Unfortunately, there is no guarantee that the electronic control, which can be continuously monitored and modified by a sound engineer, can deliver the musical experience intended by the orchestra, soloists, and conductor. In examples of say operetta, where head microphones are frequently in use, there will be a sound engineer balancing and changing volume or timbre to suit the theatre. The use of personal microphones means that there is individual sound control of each singer, both in volume and frequency response (i.e. the timbre of the note). While the objectives of such interference in the musical sound may be sensible, but there is no guarantee that the musicians and engineers are totally in agreement, nor even that they understand how much change has been introduced. For example, in a duet, the artistic control of the musical balance and intensity has moved from the performers to the sound engineer. I have heard examples where the changes were so obvious that it undermined the music. I have also realized that electronics was in use in several reputable concert halls and opera houses because of electronic problems that revealed their usage.

My concern and emphatic criticism is because I am discussing a musical change which is irreversible. It happens during a live performance. This is a different situation from making a recording, as for the recording the various protagonists of conductor, singers, and sound engineer can discuss and compare the alternative options of intensity and tonal balance (and/or to correct out-of-tune notes etc.).

Where Are the Best Seats?

This is a remarkably difficult question to answer, not least as it depends on the type of music and the instruments involved. It will also depend on whether we are discussing live performances, broadcast, or recorded music. It is going to be impossible to give a definitive answer on the position of the 'best', as we need to compromise between many competing factors (including our own sound preferences). Despite this, I will try to offer some guidelines. In all cases, we need a room or concert hall that has the scale and reverberation time to bring out the best of the work that is being played, and definitely do not want any seat that is too distant from the stage, or that has echoes or selective frequency responses. Nor do we want the distortion of seats hidden under a balcony. We always want to be close enough to hear direct sound reaching us in less than about 50 milliseconds. This means we need to be within say 60 feet from the centre of the action. (The length of a cricket pitch might be a useful estimate.) At such distances we will see synchronization between the sounds and singers' lips, conductor's movements, and violin bowing etc. We do not want detectable time delays from different parts of the orchestra, but we want enough left/right ear variations from reflections that there is a sense of spaciousness, so a central type stalls seat is best in a large auditorium. It is a real challenge to find such ideal seats with performances by a large orchestra and choir, or in a large opera house.

For orchestras on a stage that is raised above the front of an auditorium, we also need to be far enough back up the slope of the audience seating that our seat is level with, or higher than, the stage. This should give both a good view and good acoustics. For high stages, this seat position may be in conflict with our need to be fairly close. Alternatively, in a small theatre with a balcony, and stage action, as in an opera, then a front balcony seat may actually be musically better, as this offers a clear view of the stage and it may provide a better projection of the music from the orchestral pit; indeed, we may be able to see (and directly hear) the musicians hidden below the stage level.

Since many ceiling reflections can be quite delayed, and/or strengthened by focusing, we have two options. The first is to have a seat that minimizes their effects, but invariably this is a high price seat, or the other option is to accept some distance from the stage, if this is compensated by strong and clear ceiling reflected sound. I have an opera-going friend who claimed the best sound in Covent Garden was in low priced seats at the front of the upper gallery. The reason is clear, the ceiling reflection boosts the sound intensity, and the path difference between direct and reflected sound is small, so it is ideal for a strong and clear sound. Nevertheless, the pleasure of music is greatly enhanced when we have a feeling of immediacy and participation in the performance. This is precisely the intangible quality that can never be matched from a recording or broadcast, and it is a key reason why concerts can seem so much better. There is also the adrenalin effect for the performers, that in a live concert it is easy to make mistakes. These can be simple errors of playing or singing a wrong note, of not being totally in tune, or of failing to offer the phrasing and dynamics that had carefully been practised. Such errors are invariably removed on a CD version, where retakes and patching allow for corrections, but equally the multiple retakes remove the edge which is the excitement from the stress of the live delivery.

Proximity is a key factor to produce a deep response to music. I have been to local operas from good amateur groups, as well as to professional rehearsal rooms, where in both cases one is effectively on the set, and the result is incredibly powerful compared with the more distant seating. This 'participation' or 'presence' in the opera can totally outweigh the differences between a good amateur and a distant professional performance; in the case of professional rehearsal rooms the overall effect can be magical. Nevertheless, merely being close to the opera stage is no guarantee either of a good acoustic or of the best sightline. A common complaint for many of the horseshoe-shaped opera houses with seating in boxes stacked from floor to ceiling is that the boxes are ideal for the socially extrovert, but that sightline and musical quality are not perfect. I remember sitting in such a box in Vienna with an excellent view of the audience, and they of me, but with a restricted view of the deeper recesses of the stage. In one case, I heard the last part of Aida without being able to see any of the main characters, as they were in a tomb construction at the back of the deep stage, which was blocked from the higher level of the expensive box at the side of a distinctly famous opera house.

If we remove the visual link and proximity then there is a real sense of loss and this is why it is virtually impossible for a CD to generate the same depth of emotive involvement and concentration response as for a live performance. At a first guess we might imagine a video or DVD would help to redress this sense of involvement, but in my own experience it is often far worse. In video and live TV broadcasts the camera images are chosen by someone else, and so our attention is focused not by our choice but by an editor directing the camera. The images are deliberately picked to offer some highlight, for example the keyboard of a piano, a soloist, or a particularly attractive or distinctive member of the orchestra or choir. We are forced to focus on these images and simultaneously lose concentration on the music. It is entertainment but not a better musical experience.

On a more general level the trend of the younger generations to be persistently in touch with one another via mobile phones, Twitter, Facebook etc. is an attempt to reach the same immersion and contact that we want for good musical presence. It fails for exactly the same reasons as for the CD and broadcast route to listening. Electronic communication is a poor substitute for actual presence and the immediacy of real contact. Unfortunately, I suspect that many people have now passed the point that they still sense what is lacking from the indirect electronic communication, and they are now desensitized to real intimate contact. Electronics uses semiconductor materials, and I suspect that it delivers semi-communication, semi-involvement, and semi-satisfaction. I mention this as it has a severe implication for future musical events, both live and transmitted. Maybe I am in a minority, but I have seen people in a live concert viewing and recording the action on their iPad or phone (which is probably illegal). Hence, they disturbed the people in neighbouring seats, they missed the actual view of the concert because they were concentrating on the recording, and were using a sound system of minimal quality. So their subsequent home replay will be poor for both sight and sound. The 'mobile phone' generation demand the gimmicks of signal processing and imagery that I believe detracts from the music, but as a majority it will mean they are heeded, and this will be the direction taken by the CD and DVD in future variants of these systems manufacturers.

Musical Changes with Distance

Optimum distance from the performers varies with different types of instrumental music. In string quartets, close proximity is essential and actually playing in a quartet feels extremely rewarding. It is unmatched by any audience seating, except perhaps in the first one or two rows in an intimate seating arrangement. Surprisingly, being part of the action is far less attractive when playing in a larger orchestra. As a violinist one strongly hears the other member of the same desk, and the immediately surrounding instruments, but often get a far poorer impression of the total work compared with a member of the audience. There is one benefit in that, from the front desks of the orchestra, one has strong sound from the soloists, whereas the solo parts can be lost to the audience in the context of one soloist and a full orchestra. Choir music has similar benefits for the singers in very small groups, but a poor overview of the sound when singing in a large choir. The pattern appears to be that when there are many performers playing together the moderate distance audience seats are preferable but in intimate pieces close proximity is better, and for trios and quartets the best place is to be one of the players.

Seats at a similar distant in the stalls may be in the same price range, but the orchestral tone will vary across the width of the theatre. I have mentioned this before in terms of directionality of instruments and orchestral layout but to underline the effects when choosing a seat, I will consider three seats. One on the left, one in the centre, and one on the right of the hall in the same row, and then look at differences in sound from a trumpet, a clarinet, and a French horn. In my orchestral seating plans these are typically to the right, centre, and left stage. The trumpet and clarinet will basically face the conductor and the French horn will mostly be entertaining the left-side seats with higher notes emitted from the bell. If we look at the recorded patterns of the scientific measurements of the sound intensity with direction and note (Figures 14.2 and 14.3), we immediately see that there is likely to be a dramatic effect on the music and on the dynamic balance between the instruments in terms of the direct sound. My starting point is to assume that most players will face towards the conductor, so for the centrally placed clarinet, the mid-stalls seat is very acceptable, but even along the same stalls row the clarinet power will fall off by a factor of 10 by the end of the row because of the radiation pattern. The brass examples are slightly more dramatic and, if we assume that the French horns and trumpets are on opposite sides of the stage then, for audience seats at opposite ends of a row in say the centre stalls, one seat will favour the horns and the other the trumpets. Assuming that in the middle of the row the two instruments sounded equally loud, by the ends of the rows their relative intensities may differ by at least 10:1 and 1:10 compared with the middle seat. This means there is around a 100 times difference in initial intensity between the end of row seats! Even this could be a distinct underestimate for some seats, because the horn bell is directed away from the conductor off to the left-hand side of the stage, so at equal distances from the horn the power can easily vary by 100 to 1. Overall, for comparably priced seats, the relative power of these instruments will differ over a range of at least 100:1 in terms of sound intensity that comes directly from the stage. Using the French horn as an example may be an extreme case but the various instruments with frequency dependent sound patterns will definitely mean that no two people in the auditorium can ever hear the same musical balance between the intensity of the notes that are being played by each and every member of the orchestra and/or chorus.

One obvious question is that, if the power can differ by as much as 100 times or more, then why are people so tolerant of these differences? I suspect there are two main reasons. The first is that, because we use a logarithmic intensity scale, then 100 times in power is only a factor of four in the way we recognize relative loudness. The other feature is that we can never hear both versions of the sound at the same time, so effectively we are ignorant of the differences. In this situation ignorance is bliss.

The message here is that if you want to particularly enjoy the sound of certain instruments then seating is critical. Less obvious is that you must consider the sound pattern, and not just buy a seat on the side closest to your favourite instrument. An aficionado of trumpet music may actively seek a seat in the front of a right-hand balcony of a horseshoe auditorium. It will give a close view of the side of the trumpeters. Unfortunately, for all the high notes, the seat will be very inferior and a far better option would be a seat on the left of the auditorium aligned through the conductor to the trumpets. For high notes the power level will be nearly one thousand times more powerful.

I have focused on this initial sound that comes directly to us from the stage because I have said that this is the signal we intuitively use to locate the source of sound. This is only part of the music we will hear as in a concert hall as there will be reflections from the walls and ceiling, and they in turn contribute to the total power of the note as well as adding blurring and confusion between notes because of the differences in arrival time.

Once we include the reflections and so on, we have further difficulties because there is no universal criterion for defining a good seat. Every concert hall will differ and the optimum seat selection will be influenced by the type of music. Experts in such acoustics will state that the differences between sound qualities in different seats of the same hall can be as great as differences between different halls.

The Final View on Choosing a Seat

I have indicated some factors that define the musical quality and balance of instrumental input as the result of hall acoustics and orchestral layout. This will partially help us in choosing a reasonably good seating position, especially if we know the intended orchestral layout. Overall there is certainly some benefit in being central and having a seat with a good visual and acoustic positioning, but in reality we need to be familiar with the acoustics of a particular hall before we can make a proper choice. For those of us who regularly visit the same concert hall, the pattern of seat preference can become quite clear if there are events with many empty seats. The regulars will have taken the better seating. In the city where I live, one large hall includes many seats on the sides, as it is an almost circular auditorium with the stage quite far forward. This causes problems of time delays across the stage. Further, the back stalls have a muffled sound because they are underneath a low balcony. Overall this favours the middle stalls, and the front part of the balcony. The same pattern is matched by the pricing, as the management has understood the variability and demand within their hall. The musical weaknesses are inevitable because the hall was not initially designed for concerts. It is relatively circular and was initially built to be for stables and so, relative to this starting point, the results are tolerable.

My guidance is to first consider the pricing, visit an unfashionable event to see where the regulars are seated, and then experiment. Finally, never forget that even in the very best designed buildings no two seats can ever deliver identical sounds, and live concerts cannot and should not be closely compared with music from a CD. They are very different but complementary ways of listening to music. Happy listening!

CHAPTER 15

MUSIC, EMOTIONS, AND POLITICAL INFLUENCES

Introduction

Music has always been an intensely emotive force. The effects can range from pleasure or sadness to romance and religion and to patriotism or revolution. Consequently, the ability to control the music we hear and play is such a powerful tool that it is, and always has been, exploited both by those with political or religious power, and equally by those who want change or revolution. The diversity of musical viewpoints can often parallel the social and political spectra, and divisions within a society. This poses a challenge, as even attempting to write about these aspects of music and how we are influenced by them, or to find examples of failure or success in making social change opens a minefield of reactions and prejudice. Nevertheless, this is a reality that many may never have stopped to consider, but I will attempt it. To start by considering current examples might alienate large sections of any readership. So a less challenging and less divisive, wiser option is to start with historic examples, where time and distance means that we do not feel immediate personal involvement. Hopefully this will allow a more unbiased reading. Even this is not easy, as history is invariably written by the victorious sides. Further, the majority of discussions of music (and politics) had usually been written by those with an education that gave them a favoured and polarized position in society. Therefore 'music' may have different connotations for such people, compared with the music of the majority of the population. An anomaly is immediately apparent because in terms of number of listeners, volume of recordings, public events, and broadcasts, 'popular' music involves the majority of the population. Nevertheless, I am encouraged, as I have read books about music by many current writers who are active as popular musicians and find that while their style of music may differ from my more classical interest, our perception of why it excites us is remarkably similar. This aspect of music is discussed somewhat less by musicologists, who often convey an elitist esoteric tone that ignores the popular music of the general public.

Funding and Power Bases of Music

In the perspective of this final chapter I will inevitably repeat some of the comments in previous chapters, but here I am mainly concerned with the political aspects of the historical or social events. Virtually every major empire or nation has been dominated by a small ruling class where the bulk of the population were subservient, either as slaves or peasants. Modern systems may well have moved away from emperors, kings and dictators, or religion, but there is still always an elite that has most of the power over all aspects of our lives. Various estimates claim 10 per cent of the population have always controlled 90 per cent of the wealth. Modern assessments say that the wealth of the richest 50 individuals exceeds the total wealth of the poorer half of the 8 billion world population! Music has been equally dominated by the politics of each generation. Of the composers we can instantly name from the 17th century, such as Bach, Buxtehude, or Telemann, we realize their musical output was often written to suit their employers who needed weekly church items or specialist items for the nobility. Other writing did not always produce payment. Bach was never paid for his famous Brandenburg concertos. Additional funding was also feasible, and one successful exponent was Handel. Although he had patronage he also changed the style of his music to suit currently evolving musical tastes for his paying audiences. By the 18th century the church contributions had reduced, but nobility and court music could offer not just a salary but also access to a small orchestra. Haydn was in such a favoured situation (but trapped within it). His small Esterhazy band allowed him to experiment with quartets and symphonies. In his later years his fame and reputation had risen, and in concert tours to London he suddenly became financially successful. Nevertheless, for the majority of composers and performers music was not a well-paid prestigious career, and socially worse was that the majority were considered to be in a servant class. We tend to forget this, as we remember the winners such as Beethoven, Paganini, Liszt, and Brahms, who were the superstar musicians of their day.

By the 20th and 21st centuries the funding power base is driven by music that is fashionable, what music is played and recorded, and even whether recordings are marketed, broadcast, and distributed. The power brokers are rarely the musicians themselves, but instead are the agents, promoters, and recording companies. Such situations are never totally fixed, and the most recent technological additions to the music industry will blur these distinctions slightly, as music can now be widely dispersed by internet electronic communications. This offers marketing potential with worldwide access, even if the home country does not welcome it.

Responses to Music

Neuroscientists study how our brains react to many different types of stimuli, and music emerges high on the list of powerful factors. This should not be surprising as for other animals there are equally obvious responses. Birdsong is regional in language, whales 'sing', and when one whale starts a new tune the others of the pod will learn and sing it as well. In agriculture, soothing background music increases milk yield from cows and maintains a more peaceful egg producing output from hens (Mozart is claimed to be effective, but so is most calming music).

Animal responses to music are often cited. Even in my own experience I recall one dog that would go to sit with his head on the piano keyboard when my father was playing, and a cat that went out when I played the violin but returned as soon as I stopped. My friends said she had musical taste.

Humans are equally responsive, and during World War II the BBC broadcast 'Music while you work' to improve the output of factories. The music was never challenging but it helped to maintain a pleasant emotional state and it raised productivity. While writing this book I have often had background music to feel relaxed, but I choose items that are not my top favourites; because I am only using the music as background, I do not want to undermine my enjoyment of the works which I choose when I am deliberately listening to the music. The key issue is that I am using the background music to influence and achieve a relaxed mental state. This is a familiar situation but many people have commented that they believe radio programmes that continually broadcast the same snippets of popular music are actually undermining our long-term enjoyment of music in general. Our focus is less concentrated, and the five-minute fragments also mean that we cannot concentrate on a full orchestral work. The same applies to the pop music field, where most pieces are already brief. Continued and repeated exposure undermines our enjoyment, and it just becomes background noise. Perhaps this is why pop music changes rapidly in style and popularity.

For humans, music heard in the womb can define our future musical preferences, so it is clear that brain function, emotion, and attitudes or choice are malleable, and therefore control of music, especially when there are words, is a powerful tool to influence our lives. Once we recognize that this mental power structure exists, and that it can be fundamental to defining our actions, then we will appreciate why we can exploit it (or are exploited) in every aspect from love and grief to nationalistic activities and warfare. Mental control of our thoughts, ideals, and actions is equally a route to religious and political influence. Basically, music is a multi-purpose key to unlock our minds to a particular set of values and behaviour. If we doubt this, then consider why TV advertising uses so much background music, plus familiar jingles so that we instantly recognize the product and buy it.

Perhaps more surprising is that we even respond to music sung in foreign languages, and though we may not understand the words, we are sensitive to the tone and presentation. When watching foreign films with subtitles, the drama and emotions of the actions have their impact from the 'music' of the spoken delivery, and the actual text is just an adjunct to hold our attention on the plot. In opera performances with surtitles, the sounds are critical. Indeed, for foreign operas the music alone may still be enjoyable without any knowledge of the text, although we can still guess at much of the emotional content.

A necessary caveat is that styles of speech vary, not just between countries, but also between regions and social class, so overhearing a dialogue merely by tonality and emphasis can sometimes be misleading. This is also true of subtle body language and gestures, which can actually be normal in one society but very insulting in others. Those from Western Europe who travel to Bulgaria, Greece, or parts of India, may be misled by head movements for No and Yes. The meanings for shaking or nodding of the head are reversed!

Psychology and Musical Impact

In my search to follow the interplay between music and science the initial interactions were with the repeatable parts of science, such as the technology of instrumental design and electronics. There it was easy to make precise connections with quantifiable changes. In the broader view of science, one must address the undoubted responses from music on our psychological and physiological reactions. This is trickier as these will be individualistic in detail, even though there are discernible broad patterns. One book on this subject that I like is *Why we love music* by John Powell. His examples include both familiar and more extensive examples, but in these few paragraphs, I can only cite (or plagiarize) a few of them.

No matter what genre of music we prefer, we need something that is semi-predictable, so we are not too challenged. Equally, we like novelty. Therefore, a successful work will manage both. I was surprised to read that people claim they like a piece more if they have heard repeated sections. Perhaps this explains why many long classical works include repeats. I naively had assumed it was just an easy way of making a longer piece. There are also standard patterns of having section A followed by B and finishing with a return to A. Variants of this route to repetition make us feel welcomed and comfortable with material that has been heard before. (You may have noticed that I have also deliberately done this with some comments in the earlier chapters, as it is a good teaching technique.) There are many musical examples of repeated sections, interspersed with new ideas, and they may appear in patterns such as ABACADA etc. The composers had obviously been doing their psychology homework. In the same way, works may have a theme and variations in which the tune is played, and then reshaped in different rhythmic patterns, or in major and minor variants. This is a perfect approach, as we already know the tune, so feel comfortable, but each variation adds novelty, so we have both of our key elements that we subconsciously want in the music.

We may not be aware of it, but we like a structure with basic rules for the composition, perhaps not always at the complex level of polyphony and fugues, but such rule patterns also need to exist in 'spontaneous' music. These were once the norm for showpiece cadenzas in concertos at live events and are still a key element of the modern equivalent of jazz. For jazz the spontaneity is exciting, and probably more challenging than for a solo instrumentalist, as the hidden pattern of rules needs to be intuitive to all the members of the group.

Personally, I think it is sad that we no longer hear classical concerto cadenzas that are truly spontaneous and unrehearsed. They would add genuine excitement, as well as adrenalin for the performer. This point was made remarkably clearly to me in a recent concert where, as an encore to a Ravel piano concerto, the (excellent)pianist asked for a short musical phrase from the audience. She then extemporized and developed an item that progressed from a fugue to rag-time. The applause and audience response were greater than I have heard for many years.

We are so familiar with normal speech that we tend to forget the extreme flexibility that we add in any delivery, and just how much we learn from this, as well as to the actual words. We use changes in speed, pitch, volume, emphasis, or pauses,in every conversation. If we do not, then any audience will stop listening. Successful music must emulate all of these features. A conductor or group who fix a totally rigid speed and sound intensity would be rapidly unemployed. This is one reason why live music is great, recorded music is fine on the first listening, but progressively less so on hearing the same CD many times. This act of imposing our personality and meaning by all the non-verbal factors totally underlines why computer-generated music can seem unemotional and unexciting. A caveat is that with high-level software programming skills computer sounds can be modified to approach that of real music. Across all types of music our ability to make the music sing is the strength and emotional power that we enjoy. We may do this by intensity control or by bending the timing or tuning, but if it is well-performed the methods are hidden, and only the overall result is appreciated.

If you doubt our ability to recognize verbal subtlety then just consider how many ways you could say 'I love you'. If you cannot manage at least ten, then practice is urgently required.

In a film or TV programme background music of all the foregoing rules can be ignored, because the music is intended to set our mood via subtle and preferably subconscious conditioning. We sense the mood of a future scene merely from the music, or other sounds. This was even true for the silent movies where it was normal to have a few people playing mood-setting music. Their range was limited but there would always be something similar for love scenes, or for cowboys, comedy, or horror, etc.

Modern films and TV have a greatly developed art in such matters and can switch our anticipation of what is about to happen from relaxed and happy to terrified or shocked. Often this mood inducing 'music', that in the film setting is highly effective, can seem pointless or boring if heard in isolation. Indeed, it is often deliberately out of tune, or lacking a tune that we could sing. Background music is also used in many TV documentaries and science programmes. In these we want to focus on the content, and not be distracted by mood-setting music. I know I am definitely not alone in finding the background music to such science programmes both annoying and irritating.

Our sensitivity to background music, which is frequently at the subconscious level, is exploited in many areas. Perfect background music in stores or elevators should be felt but not consciously heard. Soothing music is effective in reducing crime in many environments. In supermarkets, it controls how fast we move through the shop, and therefore how much we buy, or even influences which products. I have read that playing music in a section selling wines can control which types are purchased. On days where there is German music being played in the wine section, the German wine sales increase; similarly the experiments show that the same store can increase French, Italian, or Spanish wine sale by playing background music from those countries. We are subconsciously easily controlled.

With advertisements, the background jingles alter how we perceive the advertising. In films the 'product' may be identifying the hero or villain, and so for many years a little musical phrase was linked to a specific character. These range from the 'leitmotif' of Wagnerian operas and many modern musicals to menacing sounds of the shark in Jaws, or a snippet of music which distinguishes the hero from the others.

Many people will claim that music in major keys tend to be positive and happy, whereas minor keys are sad (a major chord of C is C, E, G, C, whereas the minor one lowers the third, the E, to E flat). Sometimes this is true, but it is a cultural feature, and in some countries jolly happy tunes are in minor keys. Western examples of this alternative include Bulgaria and Spain. In many operas, 'jolly' tunes in a minor key are supposed to convey several layers of meaning (e.g. in Rigoletto, where he must entertain but is deeply distressed).

Some musical tricks to influence our subconscious are obvious and universal, others are not. I have not attempted to discuss why only some rhythmic patterns have emotional content. Nor can I quantify why the tonal quality of some singers seems fantastic, but other excellent performers have no impact at all, and particularly why the opposite view will be held by other people. There are also generation-dependent preferences that can be heard via recordings. The 19th century audience was amenable to big changes in tempi. Their violin playing discouraged use of vibrato (a frequency wobbling on a note), whereas any performer who did not use it now would be described as having a dead sound. I wonder if there is some technological reason that produced this switch in viewpoint, such as the replacement of thin gut violin strings to heavier bound ones, or if vibrato helped to disguise the poor sound quality of earlier recordings. In singing, vibrato was, and is, totally fashionable in love songs but rare in pop music (maybe autotuning kills it).

To summarize, anyone reading this book will already appreciate the emotional impact of music and its ability to stir a wide range of emotional responses, from pleasure to love or fear. Experimental psychology has added another dimension to our understanding by monitoring physical and emotional changes that we did not know were happening. The direct access and influence on our subconscious behaviour reveals that our actions can be driven by subtle background musical inputs. Rather than being masters of our own actions we can be manipulated to a considerable extent. We should not be complacent, as very similar subconscious conditioning can be introduced by the written word. Indeed, even I am trying to influence and/or widen your perception of music.

Political Influence and Control

While music is highly effective in both religion and the politics of our thinking, how this happens is unclear to most of us, although the reality is certain. Understanding the mechanisms and quantifying how music can influence our thoughts is difficult. There are broad patterns in social responses but in reality even with the best medical and sociological evidence, our behaviour can only be specified within an imprecise framework. Music is an integral factor in many religions but the usage varies considerably. With my underlying assumption that it can be exploited to help direct or control the masses it seems that understanding the role of music is as important in religion as in the 20th century political examples cited below. Even in the most simplistic and ancient religious ceremonies music and chanting have always had a place, as repetition of key phrases in the presence of the other members of the group will imprint the words in our subconscious. Once it is there we will believe them and think they are our own thoughts. This is precisely why religious (and political) training has always concentrated on the very young who are most susceptible to such control.

Not only can the repetitive factor of imprinting and remembering phrases be linked to music, but the effects can be enhanced by various mechanical acts. So, saying mantras, prayers or religious songs, while counting on beads, a rosary, or other tangible objects, helps with the retention of the ideas. Similarly, doing so in a large group, preferably at a loud volume, will link the entire group, bond them together, and eventually they unquestioningly believe the words that are being used. This is absolutely standard practice in military camps across the entire world, and it is used by groups who we may see as friendly allies, or the enemy, or as terrorists. It clearly works. Precisely the same approach is obvious in many church services I have attended. It is further enhanced if not only are the phrases brief and repetitive but they are amplified by the congregation waving hands or doing synchronized dance actions.

When young I had daily prayers at school, but these were far less effective in imprinting the teaching, compared with the singing of hymns etc. with fixed words, simple tunes, and the feeling of community with a large congregation. The school prayers have long since faded from my memory, but the hymn tunes, and even the words, can be recalled more easily. By contrast, the religious teachings of most sermons had minimal impact, and indeed often sowed major doubts, as one had the individual freedom to think and analyse what was being said. For effective domination of a population, allowing free thinking is clearly undesirable, but delivering a message via music does work. Hence music is an entry point into our subconscious that is universally applied.

Religions may have different objectives from governments, so it is not surprising that religious music has developed in different ways from secular music. Therefore, I will consider the two separately. Government control of music is not new and is obvious from historical documents. An early example comes from China around 2700 BC, where not only did the Emperor write music and encourage the generation of a musical notation, with notes named in terms of different levels of society, but also there was a Ministry devoted to defining what music was acceptable. The notation aspect was real progress, and the control was symptomatic that they were astute, and realized the emotive power of music.

Music works as a tool to bond a group of people to act together, and to say and to think the same things. A familiar example is evident at any modern football match, where supporters will sing a song to spur on their side, and as mentioned, military training includes soldiers being actively taught songs that they sing while marching, in order to strengthen their mental links and form a determination and camaraderie to act together.

The military examples stretch way back into history. While we cannot have any recordings of early musical instruments used to raise the spirits and determination for battles, there are Assyrian and Egyptian carvings (~2000 BC or earlier) showing many brass instruments that were meant to inspire, and/or broadcast commands for troop movements. Bagpipes are another later example used to intimidate the enemy, and bugle calls broadcast the orders to troops during a battle. These same techniques were used by most militia, even into the last century.

Away from the battlefield, music is politically used to unite or add jingoistic loyalty to a cause and country. In the UK, we no longer sing the National Anthem at the end of every cinema or public performance but the spirit is still entrenched and it emerges at the last night of the Proms concerts, where patriotic songs and flag waving can produce euphoria and unite the entire audience. I am told by foreign visitors that they are temporarily equally united.

The message to governments is that music needs to be written to generate supportive music that is memorable, enjoyable, easy to sing, and instantly inspires patriotism. For me a prime example must be La Marseillaise. It has all the right ingredients. It was written at the time of the French revolution. It certainly inspired successfully, as the peasant end of the spectrum overthrew the aristocracy, and totally changed the face of French and world politics. A minor unfortunate footnote to this masterpiece is that the composer lost his head under the guillotine. A very unjust reward.

Another nationalistic song, the British national anthem, is from an earlier age (probably Thomas Arne in ~1740). It is far less inspiring in terms of words, but the tune has the correct emotive appeal, and in the USA, it was temporarily used as their anthem with words of 'My country 'tis of thee'. The new lyric was by Samuel Smith in 1831. A popular anti-Jewish song written for the Nazi Fascists youth was so effective and offensive that it is still banned in modern Germany. There is a parallel here with the La Marseillaise in that the composer (Horst Wessel) was shot when still only 23.

These examples of musical power offer a warning to any government, that revolution or political change can be carried forward more effectively by popular music that focuses on the emotional psyche of the nation. There is therefore a dilemma in deciding how to inhibit such music, but to do so in a way that is not going to exacerbate anti-government feeling. In countries with elected governments one probably never consciously thinks about this. Indeed, most of a well-conditioned populace will claim that they live in a 'democracy' where the voice of the entire country is represented (both politically and musically). This may be true to some extent as it can avoid dictatorships. However, the terminology and ideal-istic image is unfortunate, as reference to the ancient 'democracies' of Greece or Rome overlooks the fact that their voting populations were a small fraction of each country, plus they excluded women and slaves. To offer some perspective, one notes that in Rome there were around 30 slaves per free Roman citizen. In reality, the balance is probably no dif-ferent today in the UK, except that the 30 are now called the working class, or include lowly paid labour in foreign countries who provide us with cheap imports. This is often slavery by another name.

Similarly, even within a modern 'democratically' elected government the leaders are normally drawn from a very narrow fraction of the total population. For example, in the UK since ~1751 there have been more than 50 Prime Ministers, but some 80 per cent have moved from public schools to political educations in either Oxford or Cambridge. Undoubtedly, they had a good education, but they and their teachers are very limited in their perspectives of the needs of the majority of the population. So even with the best intentions, they cannot fully appreciate the thinking and desires of the majority of the country. A different situation can happen in the USA where (at least in 2018) the inner cabinet of presidential advi-sors may all be billionaires.

Is a Musical Profession Free from Politics?

In politics, such data on the ruling elite are easy to recognize and quan-tify but the same power structure exists less clearly in a discussion of music. Across the musical spectrum, the power base lies with those who determine the marketing, promotions, recording, and broadcasting, plus the critics, agents, and commentators. There may be more diversity and a rapid turnover of fashionable groups and pop stars than in the non-classical areas but the same pattern is in place. For those initially too far removed from the official musical scene (e.g. hip-hop), they generated a completely separated marketing culture. The advantage of control is that, financially, the profits from the music can be far greater than those of the musicians.

Many writers and musicians have discussed the inherent problem that in such a highly competitive environment there are just a very few top name stars with immense income, and thousands who are barely able to survive. In the popular music sector the details have varied as the sales of CDs have dropped because of alternative marketing strategies. These include electronic downloads, streaming, and YouTube routes, plus there are different success levels for albums and individual songs. As of 2017, there had been a threefold increase in streaming compared with 2014. Worldwide album sales in 2016 were around 100 million, split between 50 million CD, 44 million digital, and 6 million vinyl discs. Album numbers decreased by 2017, but the scale of streaming increased to some 200 billion songs. This sounds like an extremely lucrative market but in reality the revenue per album is around 6 or 7 cents.

There is a very steep fall-off from the hits from major groups down the pecking order of fashionable artists. For 2017 numbers in the USA, the top 10 super successful artists had album sales falling from around a million (for number one) down to 150,000 by number 10. Less fashionable music could thus drop down to tens of thousands (or fewer), even for well-known groups. For easy arithmetic in hard cash numbers, I will assume that the artists receive 10 per cent of the sale price. This is not unusual, as the rest is spread among all the other agents, production, and marketing. The 10 per cent is also fairly typical of book publishing. For simplicity, I will assume the album markets at $10 (i.e. $1 to the group). But most small groups involve perhaps five members so, per person, this is down to 20 cents each. There are many hundreds of recording groups who will feel they are making progress if they sell 10,000 albums. Possibly true, but in terms of cash, this is a mere $2000 to each of them for their shiny new album.

The pattern is broadly similar in the classical end of the music spectrum, with a few superstars and thousands of highly competent performers who are mostly invisible to the general public (i.e. identical with the rest of the world, where 10 per cent have 90 per cent of the wealth). In a chapter on motivation it is therefore reasonable to ask, why do people enter the profession? Enjoyment of playing must be essential, but for orchestral players there is a downside of incessant travel and the need to perform the same popular pieces over and over again (Beethoven, Brahms, etc.), plus pieces that musically they might find unrewarding, atonal, or just plain tuneless. For opera, I read a 2016 analysis that listed that 50 per cent of commercial performances centred on works by just four composers (Verdi, Mozart, Puccini, and Wagner). The alternative of teaching has its rewards; certainly it may reduce the continuous travelling, and in music academies or universities, one assumes the quality of the students is reasonable. For the rest, life is hard but it underlines the fact that music is so crucial to us that purely logical or economic motivation is offset by the enjoyment of the subject.

We can consider who are the potential students and future musicians. Musical skills need to be honed from an early age and this needs dedication and financial backing from parents. For the classical musicians there are major costs for good teachers, transport to and from training at top academies, purchase or rental of high-grade instruments, etc. In real terms this means lots of money. Selecting one's parents with funds to do this is an ideal choice for potential musicians. Less obvious is that exactly the same background advantages of wealthy parents and good education are applicable to many other parts of the musical spectrum. It is also true of successful pop stars. Numbers I have seen cited suggest that some 15 per cent of top ranked UK pop stars had a private education (i.e. far in excess of the national UK average).

Earlier Dissemination of Music

Stardom in music requires ability, plus promotion from within the music industry. Specifically, it needs support of the powerful record and publishing companies who can effectively dictate the types of music that are produced and widely marketed. They are the real royalty of the music world, as they can control an artist's career and do so while making far more profit than the performers. In modern musical literature, examples are cited for every type of music. This is not a new phenomenon, and the seeds of such ability to control an entire music industry of what was published and extolled are documented in England as early as the reign of Elizabeth I. Basically, before the 1580s there was negligible printed music in England, and works were on manuscript. Secular music only started to appear after the 1560s. Nevertheless, music printers in Amsterdam had shown that there were serious profits to be made. This was therefore an interesting activity for an impoverished Queen. She already raised money by granting monopolies, often for payment plus 50 per cent of the profits. For political reasons this censorship/monopoly applied to printing. There was also a mixture of patronage and tax levies on transport and imports of all key materials such as iron and other metals, wood, wines, and exotic imports.

Music publishing was no exception and in 1575 William Byrd and Thomas Tallis were granted a music printing monopoly. This was not confined to English works, so they also had control and the monopoly on imports of all foreign music. A later licence was issued to Thomas Morley in 1592. Not only were these highly profitable activities but they allowed composers such as Byrd, Morley, and Tallis to market their own works at the expense of all other composers of the time. It is therefore not surprising that Early Music aficionados frequently feature their work and, without being too cynical, these composer/publishers may have failed to market composers who would have displaced them from their pole position.

The financial incentive of such marketing control would also mean that because there were intense religious biases and persecutions (the nation was deeply divided at that time) any songs and musical works that could be interpreted as undermining or criticizing the Crown would not have been disseminated.

The Use of Music for Political Control

The Elizabethan example was financially lucrative for those in control, and those excluded were not just poor but unknown. These were unfortunate circumstances, but at least they were not automatically life-threatening, although with religious connotations things were more dangerous. By contrast, there were and are many examples where failure to match the official musical image can mean prison or execution. Jumping forward 500 years from the Elizabethans, to the last century, we see that there were some blatant examples of extreme musical control from nations as politically diverse as Russia, Germany, and the USA. The Russian revolution included leaders such as Lenin, who enjoyed the classical music of Russian composers. This had obvious nationalistic overtones which could be used to unite the nation. A similar but more aggressive objective from Stalin was to only have music and composers who would stir the soul and unite the workers and send a strong musical message of how the new regime was leading to a positive and successful future. This of course meant that a high percentage of music was automatically deemed to be unsuitable, and composers who persisted with writing it had a serious threat to their lives in terms of imprisonment or execution. The classic example of proscribed composers that most people will cite is that of Shostakovich. This initially is surprising as he had a very high profile and a worldwide reputation. He was threatened, and forced to write music that was acceptable to the leadership. From the outside world this seemed an extreme action against a musician with an international status. However, the logic is obvious. If the state could threaten and intimidate such an international figure then it was obvious to every lesser-known composer that the only choice was to follow the dictates or die. It worked for Stalin, but for the rest of the world we probably lost excellent compositions from composers who have gone into oblivion.

My second 20th century example of state control of mind and music was from a politically opposite extreme type of dictator, Hitler. In nationalistic terms, he was trying to show how German music had been powerful in the past, and any music written in his time could only be allowed if it was emphasizing the glory of the Fatherland. He therefore idolized Wagner. Many believe that he based his hand gestures on a famous conductor (Gustav Mahler). His fanaticism was particularly extreme, and directed against any composer, past or present, who happened to be Jewish. (He seems to have overlooked the fact that Mahler was Jewish, and possibly so was the father of Wagner.) Not only were Jewish musicians banned, executed (or escaped from Germany), but even the music of past composers such as Mendelssohn (with Jewish parents, but baptized as a Christian) was not allowed to be performed. The only positive feature of this oppression is that many excellent Jewish musicians, and composers, escaped and enhanced the musical culture elsewhere (e.g. the UK and the USA).

Hitler is, however, an excellent example to demonstrate the emotive and life-changing influence of music on our future. At the age of 15, he attended the Wagner opera Rienzi (a hero who had a mission to lead his people out of servitude). Hitler's friend, August Kubizek, reported that this one opera performance transformed Hitler and gave him ambition, and released a visionary power to become a world leader.

A third 20th century example where one can cite the control of music was in the USA, where there was a total split along racist lines between the music of the white and black populations. Here the example is instructive in showing the power of musical influence on the attempts to change society. Dance band music and jazz did not fit the official white view of music and was linked to ongoing prejudice caused by having had a slave culture within living memory. However, the power and excitement of much of this music, as well as the quality of the performances, meant that it was taken up by the population as a whole, and certainly slowly contributed to an improvement in race relations. The colour prejudices were not confined to popular music, and in terms of the music that was normally only affordable to the rich (i.e. opera) the first non-white performers to reach the stage of major US opera houses only appeared in the latter part of the 20th century. Fortunately for us, that was a time when there were good quality recordings on vinyl discs, and so even now we can enjoy and praise their performances as being in the top rank of such operatic singing.

Popular music in the form of music festivals and performances and records were equally active politically in promoting reforms and anti-war demonstrations (e.g. during the Vietnam war). The actual degree of success of music in making political change is hard to quantify, but it is certainly a bonding factor between those who wish to make changes to political systems. Those in the music industry are confident that their influence is significant. There is no doubt that the various political activities from the music industry, together with all the more recent electronic dissemination of ideas and demands for change, are highly effective in bringing issues to a very wide public.

Political control of music had been very obvious in the UK from the BBC. In the original philosophy for broadcasting in the 1930s, the aim was not to entertain the masses, but to educate them and help shape their thinking according to that considered suitable by the Establishment. This applied equally to music. Music programmes on the Third Programme (now Radio Three) were run by an elite who disparaged popular music. As an introduction to Elizabethan motets, or avant-garde writing, this was excellent but it totally rejected 'popular' music. A classic example after the war was a Sunday evening musical entertainment of the Grand Hotel Palm Court light orchestra. The music was tuneful, popular, and attracted a mass audience. Within the BBC there were bitter arguments to keep or close it, as it 'lowered' the musical standards. However, it had the highest audience rating of any of the programmes on that channel. Therefore, it survived for many years, against the wishes of the Controllers.

Music in Religious Contexts

This is not an ideal topic for a subsection of a chapter, as vast numbers of books have been written about it from religions across the world, and details can readily be found on web searches. Nevertheless, it is clear that music, chants, and often dancing were key elements in the religious rituals of even the earliest of our ancestors. Some examples survive in more remote areas, and other works and songs have been passed from generation to generation in the case of people taken into slavery from Africa to the Americas. Musically this is interesting as it offers a unique record back several hundred years without written records.

Many religions span the entire globe so the music performed both for pleasure and in religious activities is modified by their region. For example, Christians will enjoy music in local styles which are specific to a region. Muslim participation in music depends on the sect, and in many regions around the world it is totally proscribed in any activity (not just sacred). The fact that there is such a strong negative attitude towards music, strongly underlines that it is a powerful emotional force. Other religions involve music in a variety of ways, but for secular items there can be more unity than those concerned with moral or religious issues. Instruments can be added in many cases. Some religions (e.g. Judaism) have a singer (the cantor) who will sing or chant specific prayers in a well-established format that can be traced back over several thousand years. For cantors, there are three such standard modes but the actual words may often be improvised. Exposure to this flexibility in the music is of course reflected in many of the subsequent compositions from the congregation.

Other religions include the use of bells, drums, or solo and choir singing, with or without other musical instruments. The only clear feature is
that the intent was to preserve concepts and rituals from one generation
to the next, and imprinting the stories and ideas via music was far more
effective than by speech. One may argue that often sagas, whether
Greek, Norse, Indian, or Chinese etc. repeated legends and stories without music, but in the form of poetry or verse. However, the counter argument is that the metric nature of poetry, plus rhyming words, have key
elements of music. It differs from normal speech, and many people find
it far easier to remember than speech. Poetry can be thought of as music
in another guise.

Christian religious music has had a chequered past. Initially it would
have related to the Jewish ancestry of the early Christians, but with a
move to Rome as the centre of the church, the official church language
became Latin. Music over these centuries evolved in style. There was
then a backlash against musical developments, which were perhaps too
secular, or happy, and the Tridentine Council (from 1545 to 1563) slowly
debated whether to totally ban music in churches or to allow a limited
style. The total music ban was lifted as a result of church music written
by Palestrina. This was not the end point, as later attempts were made to
ban female voices in churches.

There were the inevitable religious and political schisms within such
powerful, rich, and widespread organizations. For music, one consequence was that the Eastern Orthodox church used no instruments. A
positive musical outcome is that some superb bass singers have emerged
from Bulgaria and Russia, as their role was to replace the sound of
organs. In the USA there was apparently a problem in finding such bass
singers, and even their Orthodox Church has allowed organ music.

Musical divisions were equally apparent in splits caused by Protestants
as the church services, the Bible, and music became available in the local
languages, instead of Latin. At the time of Henry VIII there were many
attempts to keep services in Latin on the grounds that it introduced a
mystery and awe to the masses, who could not read or understand Latin.
Such understanding of political power was well developed, and several
religions require their monks (and/or normal public) to pray and chant
several times a day, including the night. In hindsight one sees that one
reason is that lack of sleep and repetitive chanting is a superb brainwash-
ing technique. Addition of electronics and amplification has enhanced

the opportunities for high volume group chanting with the aim of unification of thinking and purpose. Examples of Evangelical Festival attendance of 100,000 people have appeared, and they fully utilize the high volume enabled by modern electronics and large display screens for the generation of mass participation.

Responses to Regional Variations

Political power may be centrally based, but depending on the size of the nation or empire it will weaken wherever there are variations in language or culture. This is certainly true even for our appreciation of folk music within our own country, where the only differences are in dialect. For foreign music, as defined by the use of different totalities, musical scales, and rhythmic patterns, there may be a total failure to comprehend the music. Within the UK we can be sensitive and recognize non-English written music but may still subconsciously differentiate our assessment, appreciation, or have prejudice of even close neighbour items such as those from England, Scotland, Wales, or Ireland. In orchestral and classical music, we will usually be able to estimate the period and locality of many musical scores, and enjoy them. However, most music-loving Europeans will have difficulty with both the structure and scales of more distant lands across Asia and Africa. This is not xenophobia, but instead is a total lack of background conditioning.

Europeans may not immediately feel at ease with music from the Americas that have unfamiliar rhythmic patterns, but continued exposure and a definite desire to enjoy these alternative exotic sources of music can be rewarding. This indicates that we are not permanently hard wired to reject other forms of musical sound. Far more acceptable to a wide range of backgrounds will be the music of the pop generation that spans the globe via the internet, broadcasting, discs, and films.

What Constitutes Music that We Like

Having recognized the power of music, I have tried to understand why such a universal effect on us still works for totally different styles of music. Some features I understand such as the fact that styles of music include aspects that are evocative, or deeply disturbing, discordant, or intolerable. I cited background music in films, which uses all these types to good effect, and it is very successful in the mood setting context for *all* audiences. I find this difficult to understand because when choosing music for pleasure, there is a total diversity of tastes and styles, which vary considerably within classical and folk music to all brands of pop culture. Even more complications come in every area because the popular tastes and fashions change at an incredible rate. In pop music, survival over a decade is seen as long term.

As babies we responded well to lullabies but then took different musical routes conditioned by our location, environment, schools, friends, travel, etc. Some works are universally recognized, such as the *start* of the Beethoven Fifth symphony. However, I doubt that this work would have the same mass recognition if one started with one of the other movements. A key factor may be little motifs, phrases, and rhythms that are imprinted as triggers in our brain. Really popular items are now sometimes called earworms if they are too persistent. I can see no obvious pattern in what makes different music so important. However, I recognize that I have a very varied taste in music, but the items I like will invariably have a rhythmic pattern that is not boring, and a tune that I can easily sing, or repeat silently in my brain.

Music that we like requires enough exposure to it that we can appreciate the entire work. For classical symphonies and difficult performance items, past generations would only have heard any particular piece a few times during their life. This is totally different from the modern situation where we can hear snippets of works on the radio or can listen to recordings on CDs and MP3 or watch different performers on You Tube. Familiarity certainly helps us appreciate and remember longer works. By contrast much of the pop culture has songs that are less than 5 minutes in length and one can hear them incessantly while they are in fashion, via headsets at work or when travelling. In terms of imprinting the sound, such persistent exposure is very effective but the downside is that with excessive repetition, we no longer concentrate on what we are hearing.

Particularly with classical music we need full concentration to appreciate the subtlety but can still struggle where professional musicians perform too rapidly and we cannot follow. Professionals, like top athletes, perform far faster than the general population. It may display their ability but our level of enjoyment drops, and they lose a potential market sector for their recordings. Hearing such superstar performances is still seductive and entertaining, but it can also be frustrating when we try to play the same music ourselves.

A final positive fact is that for the music that we like and the songs that we learn they persist to offer enjoyment and pleasure well into old age. Those coping with the mental difficulties of senility and Alzheimer's may not remember or recognize their friends and relations, but apparently they are often capable of singing and enjoying the music they learnt when young. I shall now go and try to learn more songs, as a preparation for the future.

FURTHER READING

In spanning topics as diverse as music and technology there is a limited literature that addresses how they overlap and interact, although considerably detailed works exist for each area in isolation. In the course of writing this book I have used many sources for inspiration, and made frequent access to articles and opinions on the internet. Those which I have particularly liked or are detailed and informative are listed below.

Internet items were too numerous to list, and not all original sites remain accessible but invariably alternative ones will exist.

General items

Blanning, T. (2010), The Triumph of Music Belknap Press, Harvard ISBN-10: 0674057090.

Byrne, David (2013) How music works Canongate, Edinburgh ISBN 9780857862525.

Donnington, R. (1965) The interpretation of early music Faber and Faber, London.

Goodall, Howard (2013) The story of music Vintage books, London ISBN 978009958717036.

Jenkins, H. (1962) Music and the European Mind Wilfrid Dunwell, London.

Lebrecht, N. (2007) Maestros, Masterpieces and Madness Allen Lane – Penguin Group, London ISBN: 9780141028514.

Mulligan, Mark. Articles and blogs on reshaping music via streaming include: https://www.hypebot.com/hypebot, https://musicindustryblog.wordpress.com/author/musicindustryblog/page/2/

Pahlen, K. (1949) Music of the World Translation by Crown Pubs, New York.

Powell, John (2016) Why we love music: From Mozart to Metallica – The Emotional Power of Beautiful Sounds John Murray, London ISBN 9781473613768.

Randall, Dave (2017) Sound system – The political power of music Pluto Press London ISBN 9780745399300

Schroeder, C. (1889) Catechism of Violin Playing Translation of the 1895 original by Augener and Co London.

Steinhauer, Kimberly, Klimek, Mary McDonald (2017) The Estill Voice Model: Theory and Translation Amazon Kindle ASIN: B071Z43QS9

The New Oxford Companion to Music, 1999 (OUP, Oxford) ISBN 0193113163.

The Oxford Companion to Musical Instruments (1992) Ed Anthony Baines, OUP, Oxford, ISBN 9780193113343.

The rough guide to classical music (2010) Ed. J. Staines and Duncan Clark, Rough Guides Ltd London ISBN: 9781848364769.

Young, Miriama (2015) Singing the body electric: The human voice and sound technology Routledge London ISBN 9780754669869.

Items with a more scientific content

Backus, J. (1970) The acoustical foundations of music John Murray, London.

Benade, A.H. (1976) Fundamentals of Musical Acoustics OUP, New York.

Eargle, J.M. (1995) Music, Sound, and Technology Springer Verlag, New York, ISBN 9781475759389.

Fletcher, N.H., Rossing, T.D. (2008) The Physics of Musical Instruments Springer, New York 9780387983745.

Hartmann, W.M. (2013) Principles of Musical Acoustics, Springer, New York, ISBN 9781461467854.

Heller, E.J. (2012) An Experimental Approach to Sound, Music and Psychoacoustics, Princeton University Press, Princeton and Oxford, ISBN 9780691148595.

Holmes, T. (2012) Electronic and experimental music: Technology, Music, and Culture Routledge, London ISBN: 9780415957823.

Katz, Bob. (2007) Mastering Audio – the art and the science Focal press, Waltham, Mass. ISBN 9780240808376.

Manzo, V.J. (2015) Foundations of Music Technology, OUP Oxford ISBN 9781475759389.

Musical Acoustics part II (1976), Ed. Carleen M. Hutchins, Dowden Hutchinson and Ross, Stroudsburg.

Olson, H.F. (1967) Music, Physics and Engineering Dover Pubs, New York.

Rigden, J.S. (1977) Physics and the sound of music Wiley, New York.

Rossing, T.D. Moore, R.F., Wheeler, P.A. (2001) The science of sound Addison Wesley, New York ISBN 0805385657.

Springer Handbook of Acoustics (2007) Springer Verlag, New York ISBN 9780387304465.

White, D.H, and White, H.E. (1980) Physics and Music: The science of musical sound, Rinehart and Winston, Philadelphia, ISBN 139780486779348.

White, P. (2005) Sound on sound book of musical technology: a survivor's guide Sanctuary Publishing, London ISBN 1860742092.

Wood, A. (1940) Acoustics Blackie, Glasgow.

Townsend. P.D. (2016) The Dark Side of Technology, OUP Oxford ISBN 9780198790532.

A BRIEF GLOSSARY OF
LESS FAMILIAR TERMS

In the text, I have tried to define phrases that may not be familiar to readers of different backgrounds. Some terms resurface a number of times and so I have listed them here.

Accidentals Intentionally written notes that do not fit the scale which is denoted

Autotune Electronic correction of inaccurate performance

Cadenza Showpiece items built into concertos, equally a fundamental part of jazz; these are now mostly previously prepared, but some are extemporized

Chords Several related notes played at the same time

Decibels Measure of sound power (loudness); it is a logarithmic scale so increasing the power by 10 to 100 times is heard by us as a doubling of the intensity

Dissonance Collection of notes that musically clash

Flat Note change downwards by a semitone

Frequency Vibrations per second (the velocity of sound divided by the wavelength)

Formant Entire set of notes in addition to the lowest note that is being sounded

Fourier transform Mathematics that breaks down a complex repetitive wave pattern into its constituent frequencies

Harmonics Frequencies that are simply related to the note being played (e.g. ×2, ×3)

Intervals 2nd, 3rd etc., e.g. in a scale of C going from C to D, or C to E etc.

Key signature Defines the musical scale that is the basis of the music

Logarithms Counting scheme where we emphasize changes by decade, rather than in simple numbers. Extremely useful when we span a very large dynamic range. In music, our hearing spans intensities from say one to one million times. We are not interested in fine detail, only relative loudness. Going from intensity 10 to 100, we hear it as a doubling. In log terms this is from 1 to 2. Log scales are also useful for frequency as we double the frequency per octave.

Loudness From very soft to very loud are marked as ppp, pp, p. mf, f, ff, fff

Major/minor e.g. a C major key uses C,D,E,F,G,A,B,C; the minor flattens the 3rd and 6th

Modulate Change key

Music Sounds you like

Partials Harmonics that may not fit into a simple mathematical pattern

Pizzicato Notes that are plucked, instead of bowed (e.g. on a violin)

Plainsong Early music where everyone sings the tune with the same notes

Polyphony Term for several notes, tunes, and words, being sung together

Overtones Alternative word for partials

Reverberation time Time taken for a very loud noise to fade away

Scales Set of notes used per octave

Sharp Note change upwards of a semitone

Sine wave Simplest smooth up and down repetitive wave pattern.